T0074222

Can Science Be Witty?

Marc-Denis Weitze
Wolfgang Chr. Goede
Wolfgang M. Heckl
Editors

Can Science Be Witty?

Science Communication Between Critique and Cabaret

Editors
Marc-Denis Weitze
TUM School of Social Sciences
and Technology
Technical University of Munich
München, Germany

Wolfgang Chr. Goede
München
München, Germany

Wolfgang M. Heckl
Direktion
Deutsches Museum
München, Germany

ISBN 978-3-662-65752-2   ISBN 978-3-662-65753-9   (eBook)
https://doi.org/10.1007/978-3-662-65753-9

This Springer imprint is published by the registered company Springer-Verlag GmbH, DE, part of Springer Nature.
The registered company address is: Heidelberger Platz 3, 14197 Berlin, Germany

Foreword

I would've wanted to like, but allowed to like I didn't dare!

This squiggly saying by the brilliant Munich comedian Karl Valentin (1882–1948) captures the theme of this book with nano-sharp precision. How often I have found it confirmed in my career as a scientist: Esteemed and highly competent colleagues who come across as easygoing and funny in everyday life literally go stale when they take to the lectern or stage, lead a seminar, or even just discuss scientific issues in front of a few people. Yet they would have had all the talents for a lively and humorous presentation of their knowledge and findings.

Too bad! Many have the best of intentions. But in front of the public, their courage folds up like a pocket knife. They fall into the often lifelong learned routine: deadly serious, dry, outright boring—safe is safe. Science can't take a joke, to laugh at it you go and hide in the basement. Really?

Rhetorical looseness has also increasingly conquered the public sphere in Germany in recent decades. Instead of dry facts, dressed up in passive constructions, nouns and nested sentences, storytelling is gaining ground. If you want your research to be remembered, you have to combine facts with emotions, find characters, and send them on heroic journeys.

You see, it's just a short step from here to humor!

In this book, you will find a captivating firework of examples and instructions for more humor in science, ignited by practitioners and professional cabaret artists, including those from the young genre of science cabaret.

The "Big Bang Theory" around the nerdy physicist Dr. Sheldon Lee Cooper has probably persuaded more young people to study physics and science than the best-made tele-colleges. And we don't remember the textbooks directly, but we do remember, for example, experiments in class that went wrong and amused us. It's old hat in terms of learning psychology that content can simply be memorized much better via an emotional connection. So, please, become more courageous to tell anecdotes, to tell stories, to tell witty bon mots!

Browse through the 22 contributions of "Can Science Be Witty?", let them inspire you to experiment, to simply interact with more fun for yourself and others about topics of science and education. The fact that humor knows no boundaries, but only many gray areas between amusement, satire, and criticism—as another great comedian, Gerhard Polt, found out—hopefully makes the topic and this book all the more exciting!

And the next time you're in Munich, pay a visit not only to the Deutsches Museum but also to the nearby Valentin Karlstadt Musäum.

General Director of   Wolfgang M. Heckl
Deutsches Museum
Munich, Germany
Spring 2020

Contents

1

To Get Started

Marc-Denis Weitze, Wolfgang Chr. Goede, and Wolfgang M. Heckl

We want to open a new chapter in science communication. Applying cabaret to research and technology. We are convinced that science and society come together better through a smile.

Science communication is facts and emotion.

M.-D. Weitze (✉)
TUM School of Social Sciences and Technology,
Technical University of Munich, Munich, Germany
e-mail: weitze@tum.de

W. C. Goede
Science Facilitation, Munich, Germany

W. M. Heckl
Deutsches Museum, Munich, Germany
e-mail: heckl@tum.de

© The Author(s), under exclusive license to Springer-Verlag GmbH,
DE, part of Springer Nature 2023
M.-D. Weitze et al. (eds.), *Can Science Be Witty?*,
https://doi.org/10.1007/978-3-662-65753-9_1

Many people care about the facts. But what about emotion, comedy and tragedy? What role do comedy, satire, cabaret, criticism and humor play in science communication? We would like to shed light on this. In doing so, opportunities and challenges of science communication (cf. Weitze and Heckl 2016) become clear, new approaches become visible. In this respect, this book is both an extension and a deepening of the 2016 volume, to which we will refer at appropriate occasions in the following.

Science Cabaret: A Vision

Entertaining and funny science has been entering stages all over the world for quite some time. For example, at science slams, in author readings or in shows. But that is not yet science cabaret.

It was May 22, 1986, when a television director at Bayerischer Rundfunk stopped the broadcast of "Scheibenwischer" ("Windshield Wiper": satirical TV show) in order to protect the Bavarian people from ridicule about nuclear energy. Today, Max Uthoff asks why cabaret as a whole is hardly censored in our time: "Either the system is so settled that criticism doesn't really itch the powerful anymore. Or we are so tame that those on top don't recognize any real hostility" (quoted from Reiser 2019, p. 25).

Like politics, science, research and technology must also face or be exposed to satirical criticism, for example along the lines of the Rhenish carnival or the traditional "Teeth-Baring" (in Bavarian dialect "Derblecken") at the Strong Beer Festival ("Starkbieranstich") on Munich's Nockherberg by the legendary Hildebrandts, Polts and Asüls, to name only a few of the famous cabaret artists making fun of top-notch political leaders gathering there. This is precisely the niche that this book aims to conquer.

TV anchorman Hanns Joachim Friedrichs is credited with the following quote: "You can tell a good journalist by the fact that he doesn't side with a cause, even a good cause." Just as Friedrichs upholds this kind of quality journalism, Ottfried Fischer upholds the particular professional ethic of a cabaret artist: "To break free from parties. To admonish and warn unflinchingly. Remain committed to the pursuit of positive world change. To use one's own imagination and power of ideas. […] Fire up debate and discourse. Educate and inspire people, and do so with attractive, modern, and artful means. Being funny, brave, quick-witted, profound, pushing boundaries and sometimes causing pain" (quoted from Reiser 2019, p. 22). Yes: this is exactly the kind of science communication we would like to see.

While Science Slams are established and still attract a large audience, it is now time to think up the future, experiment with new forms and push with them onto the stages and into the forums. Not as an adaptation of the creative Anglo-Saxon language and science world, as has been the case in the past, but with ideas of our own, a bold and finally "Made in Germany" again! Just as many science communication forums such as Science Centers were originally invented in Germany (and then re-imported from the USA and UK).

Science and technology have turned the tide repeatedly in the history of civilization. Today's innovators, interest groups behind them, the winners and losers should be targeted not only discursively but also cabaret-style, just like the great political figures and their environment. Nevertheless, the previously valid forms of entertaining and witty science retain their justification. They also find their way into this book and embed the new.

Pigor "Down with IT" (Chorus)

- Revenge for the broken promises of IT
- Your things never work, never work, never work!
- Revenge for the broken promises of IT
- For every minute a user loses trying to figure out how
- Your screwed up menu doesn't work,
- Means no, he's not too stupid, you don't get it!

Retrieved from: https://www.pigor.de/songs-a-z/

Some Favorite Examples

"Down with IT": the starring role of the cabaret artist Pigor was always on our minds and we could fervently understand his anger when writing the texts for this book with the PC as well as during the editing work.

In fact, science communication is most exciting when it does not come from within science itself. It is precisely the outside perspective that can reveal relevance and narrow-mindedness, draw connections and reveal dead ends. Why is it that so few cabaret artists have discovered science and technology for themselves?

But of course scientists themselves also make jokes about themselves and their research. This comes across as somewhat drier and more serious, and instead of thigh-slapping, perhaps only a brief twitch of the eyebrows. A few examples may illustrate this:

- On the occasion of Max Planck's 80th birthday, on 23 April 1938, physicists' colleagues performed a humorous play "Die Präzisionsbestimmung des Planck'schen Wirkungsquantums" (The Precision Determination of Planck's Quantum of Action) (cf. Hoffmann et al. 2010). In it, diaphragm movements during laughter are measured in an experiment, supposedly based on quantum processes. And from the results of the measurements, the quantum of action is determined mathematically and its value is handed over to the celebrant by a postman in the auditorium after the curtain falls.

- In a four-volume "Encyclopedia of Philosophy and Philosophy of Science" one would hardly expect a joke article – but in fact the editor has smuggled in the "discontent sentence" in volume 4 (Mittelstrass 1996), which is pure invention. In terms of content, this does indeed relevantly describe an essential drive of science. But at the very latest with the references (with titles like "Curiosity and Asceticism. Philosophy between rainbow and (rain) barrel ", op. cit., p. 437) it becomes clear that this is a prank.

- The Ig Nobel Prize is the annual satirical award to honor scientific achievements that first make people laugh and then think ("to honor achievements that first make people laugh, and then make them think," see https://www.improbable.com/).

Here it can be clearly seen that we often only understand scientist jokes with appropriate prior knowledge. (Another example: "There are only 10 kinds of people – those who read binary and those who don't", Hurley et al. 2013, p. 33.).

The worldwide conferences of Public Communication of Science and Technology (PCST) strive for more play and art in science communication. PCST (2018) in Dunedin/New Zealand, for example, also provided a stage for the musical presentation of research and science (http://wfsj.org/v2/2018/04/23/pcst2018-engage-audiences-by-hearts-and-emotions-with-facts-and-figures/). This stage can be filled with all kinds of artistic forms and expressions … including humor.

"It Has Long Been Known ..."

Speaking of the lucrative niches of science communication: The lingua franca of science worldwide is the English idiom. If you want to succeed in research, conventional communication and all avant-garde forms of communication, you have to take to the international stage. Like the science cabaret artist Vince Ebert, who has his say in this book and takes on traditional Anglo-Saxon humor. In doing so, he and other courageous pioneers tap into not only a significant cultural reservoir of humor, but also a language that may be endowed with more humorous nuances than our native idiom.

This can be seen, for example, in the effortful paraphrasing of ignorance and incompetence, the humorous content of which is open to much interpretation between seriousness and technical constraints (Forschung and Lehre 2006). Here are a few examples:

- It has long been known (I didn't pick out the original quote),
- a definite trend is evident (these data are practically meaningless),
- typical results are shown in Fig. 1.1 (This is the most beautiful graph I have),
- correct within an order of magnitude (false),
- a careful analysis of obtainable data (Three pages of notes were destroyed when I accidentally spilled a glass of beer on them),
- it is hoped that this study will stimulate further investigation in this field (I give up!).

Research on comedy is not funny per se, but it is helpful if it differentiates and explains the basic concept (between comedy and sarcasm). A handbook by Wirth (2017)

Fig. 1.1 Cartoon by the trio of editors: Test in the experimental laboratory for scientific humor – is it tingling yet? (Graphic: Marlene Heckl)

provides access to comedy research. It presents the range of formats that can come across as comical. A monograph by Hurley et al. (2013) is packed with examples and poses the question of the (evolutionary) meaning of humor.

Is this book now a handbook, a reader, analysis or criticism? Perhaps a little of everything. It was and is important to us to collect internal and external perspectives on our topic, from and with cabaret artists, experts from theory and practice, as well as national and international examples – and in doing so, as little cool peppermint breath as possible, but rather sometimes thigh-slapping. Also and especially because we are guests of a renowned scientific publishing house.

Acknowledgements The editors would never have been able to produce this book with its many colourful spots and facets on their own. We would like to take this opportunity to express our sincere thanks to all the authors whose original and exciting texts are presented on the following pages and who develop an entire panorama. In addition, there are important contributors in the second and third rows, on whom the stage spotlight is hereby directed. Christoph Uhlhaas (Munich) brought Pigor to our attention at an early stage, thus initiating the topic. Luz Obeso kindly assisted to edit the English machine-translated version. A seminar with students at the Technical University of Munich TUM on the theory and practice of humor in science and research was a fertile field for further suggestions and introductions. Last but not least, the dynamic Munich cabaret scene also provided many impulses.

The Editors

This contribution was written by the three editors (Fig. 1.1):

Dr. Marc-Denis Weitze (Fig. 1.1, right) directs Technology Communication at the Head Office of the German Academy of Science and Engineering (acatech) in Munich. After studying chemistry, physics and philosophy in Konstanz and Munich, he obtained a doctorate in chemistry (TU Munich). He then worked at the Deutsches Museum in Munich and as a science journalist. He teaches as a private lecturer for science communication at the TU Munich and is working on a career as a science cabaret artist.

Wolfgang Chr. Goede (Fig. 1.1, left) is an international science journalist. He lives in Munich and Medellín, works as author, lecturer, facilitator around scientific and technological hotspots of our time. He studied political science and communication (MA) at the LMU Munich and was a scholarship holder of the Robert-Bosch-Stiftung in the science journalism program. His focus is the democratic and socio-political relation of science and the conflicts therein.

Wolfgang M. Heckl (Fig. 1.1, centre) is Professor of Experimental Physics and conducts research in the field of nanoscience and science communication. He is General Director of the Deutsches Museum and holds the Oskar-von-Miller Chair for Science Communication at the TU Munich School of Education. He is the author/co-author of nearly 200 original (peer reviewed) publications and nearly 150 other publications. As a member of numerous national and international committees, Heckl also advises the European Commission and the German government on nanotechnology and science communication.

References

Forschung & Lehre (ohne Autor) (2006) 7/6:424

Hoffmann D, Rößler H, Reuther G (2010) "Lachkabinett" und "großes Fest" der Physiker. Walter Grotrians "physikalischer Einakter" zu Max Plancks 80. Geburtstag. Ber. Wissenschaftsgesch. 33:30–53. https://doi.org/10.1002/bewi.201001404

Hurley M et al (2013) Inside jokes. Using humor to reverse-engineer the mind. The MIT Press (paperback edition), Cambridge

Mittelstrass J (1996) Unzufriedenheitssatz. In: Mittelstrass J (Hrsg) Enzyklopädie Philosophie und Wissenschaftstheorie. Metzler, Stuttgart, S 436–437

PCST (2018). http://wfsj.org/v2/2018/04/23/pcst2018-engage-audiences-by-hearts-and-emotions-with-facts-and-figures/. Accessed: 23 Sept 2019

Reiser W (2019) Witzischkeit und ihre Grenzen. Cicero 03(2019):14–25

Weitze MD, Heckl WM (2016) Wissenschaftskommunikation – Schlüsselideen, Akteure, Fallbeispiele. Springer, Heidelberg

Wirth U (Hrsg) (2017) Komik. Ein interdisziplinäres Handbuch. Springer, Heidelberg

2

Science Slam About Sheep Cheese and Car Tires

Alex Dreppec

Entertaining short talks on science topics – what sounds simple is the Higher School of science communication. Humor acts here as a door opener and charming invitation to those who otherwise find no access to science and technology. The inventor of the science slam sheds light on the origins and background of a genre that has brought funny science to the auditorium and the pub alike in recent years.

What Is That, Where Did That Come From?

Science Slam – does this mean that we put a colourful wig on Max Planck and he has to sing "A bit of fun is a must" on stage? Not quite. In any case, I would advise against it amicably.

A. Dreppec (✉)
Moderator and Author, Roßdorf, Germany

© The Author(s), under exclusive license to Springer-Verlag GmbH, DE, part of Springer Nature 2023
M.-D. Weitze et al. (eds.), *Can Science Be Witty?*,
https://doi.org/10.1007/978-3-662-65753-9_2

Science Slams are short presentation tournaments along the lines of poetry slams, i.e. with a time limit (ten minutes) and winners chosen by the audience, but with scientific content. In contrast to the poetry slam, anything that supports the presentation, e.g. PowerPoint, is permitted. The atmosphere should be more like that of poetry slams than that of a conference. This means that heckling is effectively rare but allowed, that emotional reactions are desired and that the audience has a certain freedom of movement. The latter also contributes to more relaxed facial features for some.

The audience chooses a winner by applause vote or in another, but as playful as possible (more on this later) – here again the reference to the poetry slam. In 2000/2001 I was on the road as a poetry slammer and at the same time I was preparing my doctoral thesis on the comprehensibility of scientific texts for publication under my civil name Dr. Alex Deppert (2001). In it, and already before (Deppert 1997), I had found out, among other things, that test readers are influenced in their assessment of the academic status of the author of a text by its comprehensibility, according to the motto: "The more incomprehensible, the more professional" (I have weighed the significance of such assessments elsewhere). As part of the same work, I was also looking for an idea for an applicable, practically relevant contribution to intelligibility research. A few weeks after it was finally too late to make changes to the text of the dissertation, I had the idea for a science slam. On the one hand, this moment of inspiration frustrated me, so I was downright angry about the idea. On the other hand, I was unsure about its feasibility. That's also why it took another three years until I submitted a concept to Darmstadt Marketing, and until 2006, when the idea actually found its way onto the stage in its (more or less) current form.

My doubts were not unjustified: The audience reacted extremely positively right away, but it was extremely difficult at first to persuade people (meaning researchers) to appear. I literally dreamt at night of the sentence "I'll take a look at it first". And with one science slammer, who initially turned down my request with reference to his upcoming relocation, I ended up having to help haul the stove, fridge, and washing machine in return for his appearance. In the end, however, I managed to fill the events.

Fortunately, media like the "Darmstädter Echo" reported on it immediately in 2006, and the "Frankfurter Rundschau" and "Spektrum der Wissenschaft" reported on it a little later, so that attempts by others to claim authorship were in vain.

At the time, it was in the air to transfer the slam idea to other content; for example, there were already "short film slams" in Stuttgart before the Science Slam. The miscibility of the components brought together in such cases, as with "song slams" for example, is not really surprising. A slam with short scientific presentations offers more "distance" between the original scientific context and the slam context. The decisive factor is the atmospheric change, which makes interesting or surprising things possible.

Anyway, the idea has spread from Darmstadt more or less all over the world and is still moderated in its birthplace by Axel Röthemeyer, who has also been there from the beginning, and me.

Is Everybody Allowed to Do That?

Anyone who conducts research can participate. The basis is research to which one has made one's own contribution – in other words, one's own share in what has been reported.

Particularly in the case of humanities scholars, one can also say: the basis is one's own original perspective on what has been reported. The fact that, for example, the public relations department presents the research of others "in person", i.e. in which the person giving the presentation has no part, and perhaps even in clothing printed with company logos, is not at all welcome – to say the least. However, in my view, "own research" cannot and should not be interpreted too narrowly either. I regret that some excellent science slammers were not qualified for German championships because of criteria that were interpreted too narrowly. I am pleased when students whose own research does not yet have the scope of that of – let's say – Max Planck participate. Improving production processes, for example, should also be accepted as research without further ado. Personally, I really like it when potentially more or less all of humanity could benefit from a new research approach and not just a particular company.

Can you learn to do that? There are now workshops for budding science slam stage performers. Wonderful, I also offer such workshops from time to time. But I hope that the workshops don't lead to a situation where only certain role models are copied and remote-controlled clones stagger onto the stages. So far, most of them have found their way onto the stage even without such a workshop.

The Science Slam is clearly not limited to junior scientists in my opinion. That would be too "cute" for me. It serves as one of many means of interdisciplinary communication (which has already led to new insights in my eyes), of communicating scientific findings back to the (research-funding) public, as a laboratory for comprehensibility and science communication, and perhaps also for acquiring young talent. Science Slam, however, is not a sandbox for up-and-coming scientists who may not yet be taken entirely seriously. Perhaps not everyone who has already collected scientific merits wants to go toe-to-toe with other, less renowned fellows and then compete with them on the basis of an audience judgement. That is, of course, to be accepted.

Empirically, doctoral students are disproportionately represented – and the grammatical form of the masculine is also meant here in terms of content. Female slammers are highly welcome and successful, but less often found on stages than their male colleagues. Why this is so is discussed in the scene from time to time. If a way could be found to change this in the long term, it would be nice.

Then there is the question: Is everyone allowed to organize it? Yes! When I renounced trying to collect royalties for organizing science slams, I did so, among other things, out of content and personal obligation to the inventor of the poetry slam, Marc Kelly Smith. The latter acted likewise so that everyone could implement his idea – which otherwise probably would not have gone around the world in such a way. It also occurred to me that the idea of the Science Slam would probably be taken up by students and remain more in the alternative and "low budget" realm. I didn't begrudge those engaged the few bucks they would make on it for a lot of effort … I thought. I didn't think of more or less large agencies that would make significant amounts of money from science slams (which, sure, have to be taxed, and then there's the retirement plan, the hungry dependents …). In the meantime, these too have contributed to the spread and development of the Science Slam. I still won't, can't, and will not hold up my hand. But I would be happy if the science slammers, where appreciable sums are earned, would always participate in it with a sense of proportion – and not in the form of higher prize money for individual "champions", because the hierarchies must remain flat and the competition playful. But researchers rarely have anything to give away at the beginning of their careers, and their enthusiasm is already exploited often enough.

The audience chooses a "Slam Champion". It was clear to me from the beginning that it wouldn't really be fair. Apples and oranges, no, actually sheep's cheese and car tires are compared. And some certainly agree simply because they think sociology is cooler than physics, or the material scientist smiles nicer than the Germanist with the ill-fitting sweater. But they all find a hearing and a great deal of interest and approval.

The audience is asked by the vast majority of presenters to behave respectfully and not to embarrass anyone. This is just for safety. So far the audience at Science Slams has always been positive and polite. I could still do well without the competition altogether. But it's part of the meaningful interaction between audience and stage, and maybe that's why it seems important to the audience.

How's That? Does It Have to Be Funny? Examples and Standardization Questions

Science slammers should win the audience over. For this, general comprehensibility – or comprehensibility for an interested, interdisciplinary audience without specialist training – is indispensable.

I tend to reject further standardisation than the ones mentioned so far – and I am always surprised how quickly self-appointed regulators and standardizers appear where certain freedoms exist and proclaim. For example, what a successful PowerPoint presentation should and should not look like. Freedom and creativity are crucial. The Science Slam must allow the stage people both. The great forefather poetry slam has changed in many ways, not always for the better in every respect, but it remains in motion. Who wants to dictate today how you can or will express yourself on a slam stage ten years from now, and how audiences will respond? I suppose coffin nails are also subject to numerous norms …

The following description of conditions that have been observed so far is therefore not intended to convey a normative character.

A wonderful sense of humor characterizes many science slam presentations. There are countless reasons for the use of humor: It is demonstrably a means to promote attention (cf. Kassner 2002) and against fatigue (cf. Klein 2004). This also makes it a door opener for serious content, among other things.

But more interesting (by now, to me anyway) is how the humor comes about and what kind of humor it is. It is often related to other features of the performance. Hill aptly summarizes often observable features: "Slammers often illustrate the relevance of their research area with references to everyday life, illustrate complex issues with images from the internet, explain difficult-to-understand content using metaphors and analogies …" (Hill 2015, p. 128).

The combination of all these elements with a surprise effect, for which a certain originality is required, provides the humor. André Lampe (2017), for example, describes the starting point of his efforts to explain his own research topic in a comprehensible way, based on a situation in which he had to justify himself to his "funders": "So there I sat, scrutinised by sceptical looks. The expectation hung in the air that I would explain … what my actions were in my thesis and how they justified the sums of money I had invested. The fact that respected authorities sat before me … did not make matters any easier …'What do you do all day, anyway?' my father asked …" (p. 56). Here the reference to the everyday world becomes clear as well as a certain self-irony, with which one generally looks better in front of an intelligent audience than with jokes at the expense of the less educated.

The closeness to everyday life is often also evident in the choice of words: For example, Nuria Cerdá-Esteban (2017) explains at the beginning of her talk, "Until now, we … know very little about how a cell decides to one day become a pancreas. What would make one spit digestive broth into the small intestine day and night?" (p. 11). This should not come across as forced – which, however, is rarely the case.

Direct contact with the audience has a positive effect, a "flat hierarchy between stage and hall". Audience and slammers are very often on first name terms with each other.

Everyday or even work clothes and the renunciation of suit and tie often symbolize the flat hierarchies. But authenticity is more decisive. If you are born with a tie, so to speak, you are also welcome on stage.

Revealing weaknesses and reporting failures increase tangibility. Lydia Möcklinghoff (2017), for example, reports on her (generally successful) research on the great anteater: "For six months of the year, I fight my way through the thorny bushes of the Brazilian Pantanal, am attacked by water buffalo – and ignored by the research object" (p. 96). One absolutely believes that she felt this way in between, and laughs delightedly at the unexpected openness.

Metaphors and Analogies Build Safe Bridges

Many of the speakers use metaphors and analogies, as it were, to bridge the "gap" between the subject-specific prior knowledge of a large part of the audience and what they want to convey to them. Some choose a kind of guiding metaphor (e.g. André Lampe with his "testicle cracker fish"), others a quick succession of numerous metaphors – e.g. Boris Lemmer, of whose wonderful performances several recordings can be found on YouTube (as well as those of others mentioned here). In the best case, these are original metaphors rather than conventional ones (cf. Deppert 2003; on metaphors in science communication, see also Weitze and Heckl 2016, pp. 60 ff.).

Remfort and Wöhrl (2017) make an analogy; they explain Heisenberg's uncertainty principle as follows:

"… a very good basis for discussion when the police again believe they have flashed you at 120 km/h in the 30 km/h zone, because either you were driving 120 or you were in the 30 km/h zone! Both at the same time are quantum mechanically impossible!" (p. 34). By the way, the audience here realizes that this is "just" an analogy and such a discussion with the police would have little chance of success. I'll bet my left big toe that more people actually know and remember more about the uncertainty principle after this lecture than before. Perhaps the advantage of comical metaphors and analogies is that, on the one hand, they clarify important aspects and, on the other, they are too outlandish to be seriously misunderstood. (This is true of many metaphors, by the way: Who believes that footballer Thomas Müller tears antelopes at night because he was once called the "lion of the national team"? Long live context.)

Often, "more scientific content" in the classical sense is found in alternation with the aforementioned elements. This becomes clear, for example, with Kai Kühne (2017), who presents himself on the slam stage as a comic artist on the side. Others do this too, but it is not the rule.

It often becomes satirical in connection with self-irony or the presentation of conditions and facts for which one's own research seeks to provide a remedy. The potential of the science slam as a forum for criticism of parts of the scientific community or as scientific satire in this sense has not yet been exhausted. Tobias Glufke (2017), for example, already intonates this. But there is more to come. Go ahead – stage and audience are ready!

Is This Populism? Thigh-Slapping?

There is criticism of science slams, of course. But I don't think there are many critical voices when I consider how surprisingly quickly science slams have spread. Apparently, many in the scientific community immediately understood that this format is not and never wanted to be any kind of competition to internal forms of communication.

Decisive for whether criticism of the science slam as an event format seems justified or not is, for me, the following question: How many viewers would have taken the trouble to read the original publication in question if they had not come to the science slam? See. The reverse case, that a small portion of the audience subsequently reads an original publication because of the Science Slam, is far less unlikely. Therefore, the science slam could only be accurately criticized if one could say, "It would have been better if Hans Karl had learned nothing at all about this and died stupidly in this regard than if he had learned what the science slam taught him." That would be strong stuff and probably pretty arrogant. Fortunately (for me?), I have very rarely thought such things at science slams. But their own arrogance does get to some people's hearts, and science that is (in parts) generally understandable takes away a demarcation line for those who feel the need to elevate themselves above the general public.

Who else pays for most of the research? That's right, the public! So they should also be allowed to demand to know something about the research. And understandably so, even if one has not previously studied twelve semesters of experimental theology. That can be annoying for the researcher. But that this demand has a long tradition and manifold justification, quite "incidentally" is also a democratic imperative, has been widely secured (cf. e.g. Weitze and Heckl 2016; Hill 2015, 2018; Deppert 2001). A society completely disintegrating into highly specialized disciplines that can hardly communicate with each other is (unfortunately) not a real utopia.

And then there is the matter of humor. Of course there are negative examples. Science Slam is an open forum and it is ultimately the responsibility of the presenters what they do on stage. However, my impression is that especially "highly educated" Germans often equate "laughter" with "ridiculous", which is perceived quite differently by US-Americans and Englishmen. This is often (not always!) noticeable in lectures in the classical science business. Perhaps this is the small kernel of truth in the cliché of the humorless German and a reason for the tendency towards differences between lectures and texts in the German- and English-speaking scientific community (cf. on scientific texts, e.g. Stolze and Deppert 1997; Deppert 2001). I often find modern art really good when I find it really funny. My non-German friends tend to understand that. The others think I'm making fun of it and find what I'm laughing at bad. Feature films that combine a serious subject with comedy hardly ever come from Germany, with a few notable exceptions. Lead-heavy problem films often alternate in this country, unfortunately, with completely pointless duds … but enough of that.

However, I would not like to sweep accusations of populism and "vulgarization" off the table so easily. At present, some common "popularizations" of political content are having such a negative effect on the cohesion of society that one has an involuntary impulse to refrain from any popularization. My co-host Axel Röthemeyer once put it this way in an inimitable way: "Of course this borders on populism. The stage attracts attention-seeking nuisances like the organic garbage can attracts fruit fly swarms. But proven again and again: some butterflies too! Also: of course it's entertainment, entertainment, fun. Sorry, we are influenced by the nineties, just like the idea of the Science Slam was impacted by that time, and that's just the way it was then. People didn't think so much about the fact that it would be better to tear down the stages because of the populism plague" (oral communication). Axel continues, by the way, as do I. Because at second glance the common populism is at least in parts the effect of a "de-intellectualization", which the Science Slam exactly opposes or – at least in my opinion – wants to oppose.

Lastly, I want to juxtapose my hope for what science slam can be and my memory of what non-communicative parts of scientific work can look like in two poems:

Speakers for Scintillation

Don't be held captive, be trapped
by nasal snob's big drivel crap
free this pie in the sky, this lame gush
from phrases' crumbs with the (lint) brush
Shaken off footnotes and handouts lie
around in the straw and down in the grass
while on the stage, all the messengers fly
and fill brandy in spyglass and opera glass
See scribes waver out of their encyclopedias
they become bright and magniloquent

Eight flushed thoughts will conquer all media
bound to follow the track of their writing's scent
They want to stand in the light with capacity's
quotations, they will fight opacity
where microphones borrow the spirit sound
where ideas use stages to stroll around
Speaker cables lay fuses to feed inspiration
from science to audience, to brain scintillation

During Stupid Required Reading

In the seat pads, hear some of the hidden mites sneeze
Just don't try to blow all this dust off the pages
cause it has been part of the printing for ages
You swallow the dust, but it still won't decrease
Before cooking your chair glue, oppressed is your backside
overfed with the bilgy gooseberry sauce
you deflate the part that's infected by dross
under which takes place the mating of dust mites
You can also, instead of just studying the book
and its grave leaden letters, simply first toast it
or epilate your ragged beard with a post-it
and tickle yourself at the inner ear's nook
Instead of perishing limply in gloomy halls
and harass yourself with affliction
just give your left middle finger some friction
rise it high, step outside of these goony walls
Say goodbye soon, because that's the idea:
if once the dumb circles close all around here
it will shoot out of open sphincters, I fear
in section "D", between "Derrida" and "diarrhea"

The Author

Alex Dreppec (Fig. 2.1) has published volumes of poetry in the legendary Hermit Press, chiliverlag and Ariel-Verlag, among others. In addition, there has fathered more than 300 other publications in German and English-language

Fig. 2.1 Science Slam creator Alex Dreppec on stage. (Photo: Ellen Eckhardt)

literary journals, textbooks, publications in numerous European countries, the USA, Canada, India and China. He is represented in anthologies and on CDs from Reclam, DTV to Mailart, e.g. with three texts in "Hell und Schnell", the standard reference co-edited by Robert Gernhardt on German-language humorous poetry from five centuries (circulation: over 50,000).

He is the inventor of the Science Slam, and also publishes non-fiction texts and scientific essays on a relatively regular basis.

What else? Vocational school teacher (with pleasure!) and active in the SlamBasis e. V. organisation team, which organises the "Krone Slam", among other things. Leader of writing workshops at schools and universities. Appearances on radio and television (Arte, MDR, HR1, WDR). Around 1995 release of pop music with placements in Airplay – and DJ charts. Before times Poetry-Slam-Champion in numerous cities. One of the founders of the "Darmstädter Dichterschlacht" (Darmstadt Battle of Poets), which was sold out at times with over 1000 spectators.

Doctor of psychology, doctoral scholar of the state of Hesse, inventor of a salad dressing: warm peanut butter and mix with warm water and yogurt, salt, pepper, lemon or light vinegar, a little sugar or acacia honey, coriander (fresh and finely chopped or ground). To taste a little (spring) onion, garlic does not fit. Enjoy your meal!

References

Cerdá-Esteban N (2017) Wenn ich groß bin, werde ich Bauchspeicheldrüse! In: Lampe André (Hrsg) Ein Science-Slam-Buch. Lektora, Paderborn, S 10–20

Deppert A (1997) Die Wirkung von Fachstilmerkmalen auf Leser unterschiedlicher Vorbildung. Eine empirische Untersuchung an psychologischen Fachtexten. Fachsprache 19(3–4):111–121

Deppert A (2001) Verstehen und Verständlichkeit. Wissenschaftstexte und die Rolle themaspezifischen Vorwissens. DUV, Wiesbaden

Deppert A (2003) Die Wahl der Metaphern. metaphorik.de 05/2003. https://www.metaphorik.de/sites/www.metaphorik.

de/files/journal-pdf/05_2003_deppert.pdf. Accessed: 18 Aug 2019

Glufke T (2017) Von der Leere der Subventionskunst oder warum Hunger Kunst antreiben kann. In: Lampe André (Hrsg) Ein Science-Slam-Buch. Lektora, Paderborn, S 102–113

Hill M (2015) Science Slam und die Geschichte der Kommunikation von wissenschaftlichem Wissen an außeruniversitären Öffentlichkeiten. In: Engelschalt J, Maibaum A (Hrsg) Auf der Suche nach den Tatsachen. Proceedings der 1. Tagung des Nachwuchsnetzwerks INSIST. http://insist-network.com/wp-content/uploads/2016/04/Hill-Science-Slam-Engeschalt-2016.pdf. Accessed: 18 Aug 2019

Hill M (2018) Die Versinnbildlichung von Gesellschaftswissenschaft. Herausforderung Science Slam. In: Selke S, Treibel A (Hrsg) Öffentliche Gesellschaftswissenschaften, Öffentliche Wissenschaft und gesellschaftlicher Wandel. Springer VS, Wiesbaden, S 169–186

Kassner D (2002) Lachend Unterrichtsziele erreichen. In: Gruntz-Stoll J, Rißland B (Hrsg) Lachen macht Schule. Julius Klinkhardt, Bad Heilbrunn, S 43–56

Klein G (2004) Lachen – Lernen – Schule. In: Kautter H, Munz W (Hrsg) Schule und Emotion. Universitätsverlag Winter, Heidelberg, S 121–140

Kühne K (2017) Politische Arbeitsrechtssprechung. In: Lampe A (Hrsg) Ein Science-Slam-Buch. Lektora, Paderborn, S 38–55

Lampe A (2017) Die Geschichte vom Hodenknackerfisch. In: Lampe A (Hrsg) Ein Science-Slam-Buch. Lektora, Paderborn, S 56–65

Möcklinghoff L (2017) Erbsenhirnparalleluniversumsforschung am Großen Ameisenbären. In: Lampe André (Hrsg) Ein Science-Slam-Buch. Lektora, Paderborn, S 90–101

Remfort R, Wöhrl N (2017) Diamanten aus der Mikrowelle. In: Lampe André (Hrsg) Ein Science-Slam-Buch. Lektora, Paderborn, S 22–36

Weitze MD, Heckl WM (2016) Wissenschaftskommunikation – Schlüsselideen, Akteure, Fallbeispiele. Springer, Heidelberg

3

Laughter Tears Down Walls

Vince Ebert

German science humor goes international. Vince Ebert is one of the few German cabaret artists with a scientific education and the only one so far to perform in English abroad, for example in Scotland at the Edinburgh Fringe and the USA. This here is his both serious and entertaining plea for intellectual and scientific freedom – realized through humor. Because the best way to peoples minds is via a laughing belly.

Many people think that physics is a very abstract subject. Those who do science must have a special eye for complicated formulas and dry laws. In reality, the core idea of science is very simple. Basically, scientific thinking is nothing more than a method of testing conjectures. For example, if I

V. Ebert (✉)
HERBERT Management, Frankfurt a. M, Germany
e-mail: info@vince-ebert.de

guess, "There might be some beer left in the refrigerator," and I go check, I am basically already engaging in a preliminary form of science. Thats quite different in theology. There, conjectures are not usually checked. For example, if I just assert, "There's beer in the fridge!" then I'm a theologian. If I check, I'm a scientist. If I look, find nothing, and still claim, "There's beer in it," then I'm an esotericist.

But what do I do when the refrigerator is locked? Then I must try to find out the truth in some other way. I can shake it, I can weigh it, I can X-ray it. I can even torch the darn thing and then test the products of combustion for beer. All this, of course, makes the thing extremely costly and tedious. That's why an esotericist can claim more nonsense in five minutes than a scientist can disprove in a lifetime.

But even when I've done all sorts of experiments, I never have full certainty that there's actually beer in that stupid fridge. A residual doubt always remains. That is the reason why there is no absolutely certain knowledge in science.

It's the same in normal life. A farmer comes to feed the geese every morning. The geese think to themselves: Man, our farmer is a super buddy … Shortly before Christmas, however, the geese suddenly realize: Something is rotten in our theory …

In technical jargon, this is called "falsifiability". Each theory is considered correct until it is replaced by a better one. And thereby we err upwards, so to speak!

Congratulations! You've just understood 2500 years of scientific history … With the help of two or three little gags.

Joke Logic

I've told that refrigerator example a couple hundred times on stage by now, and afterwards people have often come up to me and said, "I'm not into physics and science at all, I never understood it in school, but that example opened a door for me."

Humor breaks down the entrenched way of thinking. It changes the perspective and, in a sense, forces our brain to set up different connections. Technically, the essence of humor is that our expectations are misdirected in unexpected ways. Basically, a joke is a logical contradiction, a flaw in the system. And our brain then resolves that error into laughter. Two coliform bacteria walk into a bar. Says the bartender, "Sorry, we don't serve E. coli." "Why serve?" the two reply. "We've been working in your kitchen for weeks."

Nevertheless, in our culture – in contrast to the Anglo-American area – the idea that knowledge transfer and humor are mutually exclusive is still very widespread. But the exact opposite is the case, which is now known from brain research. The psychiatrist Manfred Spitzer wrote the beautiful sentence: "A happy brain is an eager learner."

Create Curiosity

- What was the main industry of the Afghans? Dogs and drugs.
- What do you call a person who dies for his faith? Dead.
- Name five animals that live in Africa! Three lions and two giraffes.

These very funny answers to exam questions are all from students who have a real sense of humor. And they are also extremely creative. Unfortunately, however, they are also extremely rare. That's because school tends to reward conformity, not whimsical rule-breaking. Our current education system dates back to the eighteenth and nineteenth centuries and is designed much like a steam engine: you stuff something in at the top and something comes out at the bottom. It was created to serve economic interests during the Industrial Revolution. Our curriculum today still comes from that era. A standardized one-size-fits-all menu of reading, arithmetic, and writing. Dates, vocabulary, binomial formulas. Unusual angles and rule-breaking are not in the curriculum. "Vince is very curious and often asks weird questions," my homeroom teacher told my mother at parent-teacher conferences. I think that worried her a lot at the time.

A few years ago, creativity researchers George Land and Beth Jarman had five-year-olds take a test originally developed by NASA to spot particularly innovative engineers and developers. The test was, "Find as many uses for bricks as possible."

This simple question measures divergent thinking, that is, the ability to find as many answers as possible. And this can only be done if the answers become ever more oblique, unorthodox and original. Divergent thinking undermines the rules of logical linear thinking that is mainly taught in our education system.

The five-year-olds did great across the board. 98% made it into the "highly creative" category and would get a job at NASA right away if their parents didn't think they should finish school first. For ten-year-olds, the rate was down to 30%. Adults over twenty-five were at 2%.

A good education system, in my opinion, should teach students not so much *what* to think, but *how* to think. A good education system asks questions and does not give pre-formulated answers. In my school days, there were assessments in categories such as order, diligence, cooperation and conduct. Now (and back then) other categories would be more important: creativity, originality, curiosity. And very important: a sense of humor.

Break the Rules

If you take a look at the really great natural scientists in history, you will notice that many of them conveyed their content with quasi-cabaret interludes. Nobel Prize winner Richard Feynman, for example, once said, "Anyone who does physics all his life and doesn't go crazy over it has understood nothing at all." Albert Einstein is reputed to have said, "The horizon of many people is a circle with radius zero. And they call that their point of view."

Geniuses like Erwin Schrödinger, Niels Bohr or Wolfgang Pauli all had a wit. And I will stick to it: they were so brilliant not in spite of their wit, but precisely because of it.

Because: humor breaks rules, humor is anarchistic. You, dear reader, will not receive a Nobel Prize if you think along the usual lines, only if you throw the usual rules overboard. You see something that countless colleagues have seen before you, but you think something that no one has thought before you.

How exactly comedy works in our brain is still being intensely researched to this day. At the London Institute of Neurology, test subjects were put into a brain scanner and told more or less funny jokes. When a punch line hit, it lit up particularly strongly in the so-called frontal lobe. An area in the brain responsible for reward. Interestingly, people who have suffered an injury to this same forebrain lobe due to an accident completely lose their sense of humor. These people may get the joke, but they can't laugh about it. Conversely, of course, not every humorless person automatically has brain damage. Otherwise half of East Westphalia would have to undergo neurological treatment.

All neuroscientists agree on one point: humor is an enormous intellectual ability. Because by linking things that don't actually go together, cognition magically emerges. Humor uncovers the structures beneath the surface, so to speak. Or metaphorically speaking: Laughter turns walls into windows.

Convince with Humor

A few years ago, I was invited to a gig where I was supposed to deliver a cabaret performance on a technology topic in front of Greenpeace employees. In front of me sat three hundred sceptical environmental activists with a slightly aggressive attitude towards a technophile like me. I began my show with a fictional story in which I asked a Greenpeace activist, "Why do you only ever demonstrate against fur coats, but never against leather jackets?" To which he replied: "Because its less risky to harass older ladies than the Hells Angels." To my great astonishment the gag actually worked and my audience could laugh at themselves and further on also at me and my gags. So with humor you can set a lot of things in motion.

As we all know, it is insanely difficult – if not impossible – to get scientific facts across to a person who adheres to an erroneous belief, a conspiracy theory or a pseudoscience. Which is largely due to the fact that our brains are very adept at lying through our teeth. We all like to cobble together our own truths. It's almost impossible to avoid. After all, it is easiest to deceive oneself. And we know from our own experience that you can rarely convince someone who wants to believe in something with rational arguments. If you throw in some humor your chances tend to become slightly better.

Laughter Kills Fear

Take homeopathy, for example. As is well known, this is based on the principle of high dilution. In Belladonna D30, the original substance is diluted 30 times in succession by a solvent. From the 24th dilution stage onwards there is no Belladonna molecule in the solution at all, but it is still supposed to work. This is similar to throwing a car key into the river Main in Würzburg and then trying to start the car with the Main water in Frankfurt.

When I tell this joke in my shows, I always realize to my great pleasure that even diehard homeopathy fans have to smile against their will. Of course, I realize that people who believe in inedia, horoscopes, or Bach flower remedies don't walk out of my programs saying, "My goodness, what nonsense did I believe in…" But if I can get people to laugh at their own beliefs, then maybe next time they won't be able to take them quite so seriously anymore.

In the motion picture "The Name of the Rose", the librarian Jorge de Burgos says: "Laughter kills fear, and without fear there can be no faith." That's why he killed all the monks who wanted to read the forbidden satirical "Second Book of Poetics" by Aristotle.

Satire Disenchants Taboos

It is no coincidence that the hallmark of all totalitarian rulers and regimes is their leaden humorlessness. For when the people begin to laugh at the tyrant, the tyrant loses power and the negative becomes more clearly visible. Dictatorships have always fought satire, caricatures and jokes because they elegantly expose the true circumstances. In the Third Reich, the cabaret artist Werner Finck asked an eager SS man during a performance: "Am I going too fast? Can you follow? Or do I have to follow you?"

The dividing line between a society open to the world and the future and a totalitarian one has always run along the line of humor. Churchill, after all, reportedly once said, "I collect jokes that people make about me." And Stalin is supposed to have replied, "I collect people who make jokes about me."

Humor breaks rules, is anarchistic and thus shows us unorthodox perspectives and views – and sometimes even solutions. Satire sets the record straight, exposes taboos and unspoken problems. And all this with a confident, serene smile.

I think science communicators have a duty not only to inform an already convinced specialist audience, but also to reach out to those who have nothing to do with science. Those who often don't even understand what exactly science is in the first place. Because when large parts of society adhere to pseudo-scientific mumbo-jumbo, this is not harmless fun, but it destroys everything that the fathers of the Enlightenment fought for 250 years ago.

Be Sceptical

Every few years, the magazine "Cicero" compiles a ranking list of the 500 most important intellectuals in Germany. This list is made up of those who have had a high presence in German print media over the last ten years, who have been frequently quoted on the internet and who have had many Google hits. According to the 2018 ranking, there are just two natural scientists among the top 100 intellectuals. The public discussion about energy supply, risk assessment, genetic engineering, climate change or digitalization is largely determined by humanities scholars, theologians, writers, lawyers, theatre people. But why is it taken for granted that a Catholic abbot can make a more profound contribution to stem cell research than a molecular biologist? Because monks reproduce by cell division?

A few years ago, Dietrich Schwanitz, now deceased, a professor of literature, wrote in his best-selling book "Bildung": "Knowledge of the natural sciences is taught in school; it also contributes some to the understanding of nature, but little to the understanding of culture. Scientific knowledge need not be hidden, but it does not belong to education."

A – in my opinion – very arrogant attitude. Because anyone who does natural science not only learns about formulas and numbers, but also learns how the world works, where the limits of knowledge are, and above all learns what science means: to be sceptical, to ask critical questions, not to trust authorities blindly.

Emotionalize Facts

Richard Feynman once said, "Natural science is a long story of how we learned not to kid ourselves anymore." Just 400 years ago, every storm and disease, everything that was somehow out of the ordinary, was attributed to witchcraft. Today, molecular biology and meteorology provide an explanation for what was enough to burn women just a few centuries ago. The greatest gift of science is that it teaches us something about the use of mental freedom.

And that is why we all have the duty to bring this freedom of mind to the people with all the means at our disposal. Humor is certainly not the only means. But it is a very effective one.

For almost 20 years, as a science cabaret artist, I have tried to explain scientific relationships with the laws of humor in my stage shows, TV programs, books and lectures. Because I am convinced that humor is a very powerful tool for conveying knowledge. If we want to get people excited about science, we must first succeed in emotionalizing scientific facts. Because the only way to get into people's heads is through their gut. Humor awakens these emotions. Because humor touches the emotional level.

Computers with a Sense of Humor?

Incidentally, that's also why computers don't have a sense of humor. Because they have no feelings. And because they don't make mistakes. That's why they have no sense of absurdity. In stark contrast to us humans. Our brains find anything exciting that doesn't fit the pattern. Anything that doesn't add up. Then it wakes up from its stand-by mode. We are probably way worse in doing math than a Pentium 4 processor, but we do have a sense of humor. Spotting a good friend from 60 yards behind is easy for us. A computer can't do that. It doesn't have a good friend. But it can multiply 73 by 26 in a flash. A person who can do that usually doesn't have a good friend either …

The Author

Vince Ebert (Fig. 3.1) was born in Amorbach in the Odenwald in 1968 and studied physics at the Julius Maximilian University in Würzburg. After graduating, he first worked in a management consultancy and in market research before starting his career as a cabaret artist in 1998.

Fig. 3.1 Science cabaret artist Vince Ebert rocked New York with Fine German Humor. (Photo: Frank Eidel)

His stage programs made him known as a science cabaret artist who inspires both laymen and scientific audiences with wordplay and comedy. His current show is titled "Make Science Great Again!" His books have sold over half a million copies, and some have been on bestseller lists for months. His latest book "Broadway statt Jakobsweg" (Broadway instead of St. James Way) was published by dtv in September 2020.

Vince Ebert is a regular presenter of the ARD TV programme "Wissen vor acht – Werkstatt". (Knowledge before eight – Workshop). Whether as a cabaret artist, author or speaker: Vince Ebert's concern is to present scientific contexts with the laws of humor. As a science comedian, he has also earned great acclaim abroad, at the Edinburgh Fringe Festival and in the USA.

4

"Die Anstalt" TV Show as an Example of Criticism, Satire, Humor in Science Communication

Dominik Eckert

In the winter semester 2018/2019, the Chair of Science Communication at TU München offered the seminar "Can science be funny?". Students chose case studies, presented and discussed them. Here is an example from an engineering student who analyzed a popular TV format called "The Anstalt" (The Institution).

"Die Anstalt" is a political-satirical program on the ZDF television channel that is broadcasted once a month on Tuesday evenings at 10:15 pm. The show is hosted by the two cabaret artists Max Uthoff and Claus von Wagner. The format picks up on current economic, social or political events and attempts to deal with them humorously using satire. In doing so, the sometimes highly complex issues are

D. Eckert (✉)
TUM Student, Munich, Germany
e-mail: dominik.eckert@tum.de

brought to a level that requires little prior knowledge, presented from sometimes unusual perspectives and processed in a cabaret-like manner.

"Die Anstalt" sees itself as a satirical entertainment show. Satire is an art form in which people, incidents or events are mockingly, subtly and humorously denounced. It serves to entertain, to criticize and also to teach, which makes it particularly interesting for science communication.

The two scenes discussed here are from the March 7, 2017 broadcast dealing with the "Dieselgate" emissions scandal. In the role-play or debate that can be seen, Max Uthoff slips into the role of the then CEO of Daimler AG, Dieter Zetsche, while Claus von Wagner plays the accuser of wrongdoing in the cover-up scandal.

How Is Science Communicated Here?

The Double Staging

Max Uthoff succeeds in playing the role of Daimler boss Dieter Zetsche with minimal costuming. The "Zetsche-typical" moustache and nickel glasses suffice as clear recognition features. His upright posture and routine gestures make the parody clear. He manages to be credibly perceived by the audience as the sublime car boss, naturally innocent of the whole affair.

Institutional Disagreement

The first scene deals with the dangers caused in cities by car exhaust, especially by the nitrogen oxides contained in diesel exhaust. In the course of the conversation, Claus von Wagner takes on the dominant role of the narrator and vividly and exaggeratedly describes the consequences of the "deadly" nitrogen oxides. In contrast, Dieter Zetsche, alias

Max Uthoff, tries to play down the danger, to legitimize it and to ridicule it.

The second scene deals with the manipulative defeat devices in motor vehicles. There is now a role reversal between the two cabaret artists. Now it is Max Uthoff, alias Dieter Zetsche, who comes out of the defensive and professionally reveals the functioning and the underlying concept of the defeat devices installed in cars in a highly simplified way for the audience. By changing the way the conversation is conducted, the two cabaret artists succeed in presenting both their arguments and the technical background in a credible and varied way.

Visual Representation

In order to convey the facts to the viewer as simply as possible, the facts are clearly listed on a magnetic board with the help of word and picture cards. A childlike, naïve-looking street map was chosen as the background motif. In the course of the discussion, the board is changed each time in order to visualize the complex processes in the car companies during the manipulations and to make them transparent for the audience. This is conveyed in lively role-play. Other means of presentation, such as short films, are therefore unnecessary.

Overstatement and Understatement

Various elements of satire can be detected in the scenes shown. The two cabaret artists often use the stylistic device of understatement or exaggeration. In this way, arguments are clearly emphasised, which can, however, also cause uncertainty among the audience as to the truth of the statements. The argument presented in the show is interspersed with ambiguous and ironic statements. The aim here is not the dry imparting of school knowledge. The viewers should rather be encouraged in an amusing way to question the information offered and to form their own opinion.

Fact Check

As viewers may be unsure, due to the satirical nature of the topic, which arguments only served the humorous presentation of the topic and which are based on actual facts, "Die Anstalt" offers an online "fact check" for each show (e.g. https://www.zdf.de/assets/faktencheck-am-7-maerz-100-o riginal?cb=1491342351109. Access: 19.01.2019). This comprises an approx. 30-page document in which all sources of the claims or facts prepared in the show are deposited, clearly arranged by topic and partly even commented. The paper of the former Federal Minister of Transport Alexander Dobrindt, to which the second scene refers and which explains illegal defeat devices, can also be found in excerpts in the "fact check".

"Die Anstalt": A Prime Example of Science Communication

Dieselgate or the emissions scandal is an extremely complex political and, in particular, economic issue that directly affects almost the entire public. In order to understand this, however, the technical and scientific causes and consequences must first be worked through at a level that can be understood by laypeople. This is often given too little attention in the political debate, but is taken into account here by the show.

"Die Anstalt" offers access to complex topics in a humorous way. Parody and role-playing convey the facts to the viewer from different perspectives. Satirical stylistic devices illustrate the criticism of the behavior of the car industry against the technical background. With the help of an easy-to-grasp blackboard diagram, the cabaret artists succeed in bringing the facts down to a simple level that can be understood by everyone. Orders of magnitude, such as the different amounts of exhaust gases emitted, are clearly conveyed visually with clouds of exhaust gases of different sizes. This makes it easier for the audience to understand the matter.

It is striking – especially in times of social media and "fake news" – that a cabaret show publishes a serious collection of sources accessible to everyone. This ensures transparency and perhaps encourages viewers to do further, independent research. In "Dieselgate", Max Uthoff and Claus von Wagner succeed not only in entertaining the audience with a combination of humor and vividly presented facts, but also in educating them about political and especially technical backgrounds at the same time. It is precisely this mixture of entertainment and simultaneous knowledge transfer that makes the satire as seen in "Die Anstalt" a very meaningful example for science and technology communication.

The Author

Dominik Eckert (Fig. 4.1) was born in Munich in 1996 and is currently studying engineering at the Technical University of Munich TUM.

Fig. 4.1 TUM student Dominik Eckert investigating sources of the show

5

Date an Engineer

Georg Eggers

Science-Pop is a new genre that has not yet found its way into the hit parades, but is starting to blossom on cabaret stages. Here is an example of a song text that can be heard from time to time on stages in and around Munich (in German: "Rendezvous mit einem Ingenieur").

Don't be afraid, to date an engineer,
there is no cause for worries or for fear:
His checkered shirt shines bright in white and blue,
while he is gently staring at your shoe.

G. Eggers (✉)
Hochschule München, München, Deutschland
e-mail: kontakt@physik-des-scheiterns.de

45

Say, how often, pretty lady, has your honest heart
   been broken
by a vow a sleazy sportsman or a pilot once has spoken,
so just dump those hairy playboys that will only leave
   you crying
cause an Engineer does science, you know and science is
   never lying.

So please take heart and date an engineer.
And if you think, his compliments are queer:
"Your smile is just as bright as laser light"
Then be assured: He's technically right.

Just stop hanging out in bars where all those macho jerks
   are sneering.
Go and find your Mr. Right here in the field of engineering
and don't worry if he's shy or just a little bit short-spoken
he won't jabber cheesy compliments or vows that are
   just broken.

And if you make an engineer your spouse
he'll start next day with building you a house.
And if you feel, his hunger and his thirst
You can be sure: you are his very first.

He will daily be of use whenever fuses do need mending
and his credit card will always be at hand for your
   free-spending,
you'll be living quite alone but in a land of milk and honey
cause he's working day and night but he'll be making
   pots of money.

An Engineer, means happiness and glee.
He gets you lucky with accuracy.
But if you crave incredibility
Then go and mate a physics phd.

The Author

Georg (or backwards: Groeg) Eggers (Fig. 5.1) studied physics, then shortly after the turn of the millennium took up a position as a development engineer in Munich during the day and performed in cabaret shows and poetry slams after dark. In 2011, these diversified skills qualified him for the president position (part-time) of the "Freie Universität Schwabing", which was founded in the intellectual environment of the Munich Lach- und Schießgesellschaft (cabaret theatre with a longstanding tradition) in protest against the introduction of tuition fees at Universities. Since then he teaches there "The Physics of Failure" in numerous highly experimental courses.

With the help of this "didactic pre-qualification", Georg Eggers became a professor at the Munich University of

Fig. 5.1 Groeg can do more than just math and electronics – here in dialogue with a praying mantis

Applied Sciences at the end of 2012, where he represents the subjects of signal processing, sensor technology, measurement technology and the fundamentals of electrical engineering. In order to create a professional foothold outside of logical reason in the face of the growing number of flat-earthers, vaccination opponents and climate deniers, he and his colleague Michael Sachs (mathematics) founded a singer duo that aims to make scientific worldviews emotionally accessible with titles such as "Wenn Ingenieure zu sehr lieben" (When Engineers Love Too Much) or "Seit heute Morgen hab ich deinen fiesen Schnupfen" (Since This Morning I've Got Your Nasty Sniffles). The tandem also hosts the "Applied Science Slams" at the University of Applied Sciences Munich.

6

Paradigm Disease: An Almost Incurable Scientific Epidemic

Peter L. W. Finke

Sometimes the one who thinks he is healthy is sick. And vice versa. This is also and especially true for scientists, as explained here from a theoretical perspective and with a twinkle in the eye.

Frightening Diagnosis

From time to time, a strange disease afflicts scientists who recover from it only with difficulty, only after a long convalescence, or never: paradigm disease.

This disease can be recognized by the fact that researchers and teachers in full possession of their intellectual powers abandon for an indefinite period of time that which they usually do without regard to the person and convey as

P. L. W. Finke (✉)
University of Bielefeld, Bielefeld, Germany

© The Author(s), under exclusive license to Springer-Verlag GmbH, DE, part of Springer Nature 2023
M.-D. Weitze et al. (eds.), *Can Science Be Witty?*,
https://doi.org/10.1007/978-3-662-65753-9_6

important: critical thinking. Instead, a certain view is now for them all at once considered dogma. That is why the paradigm disease is so dangerous. It seems to be almost incurable once it has broken out. In most cases, scientists were still healthy in their student days and are afflicted at the earliest when they approach their exams; sometimes even later. Doctorate and habilitation are quite dangerous phases for sustained infections, but at the latest when someone holds a professorship, he shows the typical symptoms of the disease. Only a few scientists reach old age completely healthy.

Paradigm disease is an infection with the pathogen *Paradigma disciplinensis*, which has been known for several decades and was first described by the US physics historian Thomas S. Kuhn in 1962 in his book "The Structure of Scientific Revolutions" as the causative agent (Kuhn 1970). Apparently, this is a virus that is capable of spreading epidemically under certain circumstances. Although effective almost everywhere long before, the virus in question has only been recognized as such since that publication, but until now its harmful effects have apparently been completely underestimated. To a large extent, it has even been considered a desirable companion for all kinds of research and teaching.

A scientist who shows the signs of infestation with *paradigma disciplinensis* enjoys to this day a general reputation as a powerfully and convincingly acting representative of a mature discipline that no longer wavers with regard to its basic positions and that has become textbook-worthy; in other words, he of all people was and is considered healthy. In this respect, it is indeed very surprising that the same phenomena can now suddenly be interpreted quite differently, namely as symptoms of illness.

Interestingly enough, even the first describer of Paradigma succumbed to exactly this deception of misinterpreting the actual harm as benefit. For he actually based his initial description of the virus on the aforementioned confusion of the sick with the healthy, thus characterizing a science afflicted by it as "normal" and one freed from it as in some sense "out of touch". He misinterprets the latter as undergoing a phase of "extraordinary research" in which the accustomed order of the normal is temporarily lost. This detail from its history of discovery alone shows how unusually dangerous, namely mentally disorienting, the virus is. It thus proves that the phenomena with which we are concerned are still much more dangerous than was said before. But they would be only half as great a danger if they really affected only the actors in the sciences, and not the disciplines themselves. It is, however, quite evident that this is what is happening. And this was also recognised by their discoverer, who – himself infected – mistook illness for health.

The fact that sick scientists can infect their entire discipline is very unusual in that, despite the close relationship between science and scientists, characteristics of the latter do not, according to current knowledge, transfer to science. Some sensitivities of researchers may have an influence on the circumstances under which science takes place; the research results, however, and especially their truth status, were and are still considered to be unaffected by this. Those research circumstances do influence – according to the usual view – the course of *the* history of discovery *(context of discovery),* but in no way the question of whether the results of this work are ultimately to be regarded as true or false *(context of justification).* Now we know better: *O sancta simplicitas theoriae scientiarum!*

Scientists often exhibit unusual behaviors; this has always been the case, but it has never seriously affected the dignity of science. They are not infrequently vain, power-hungry, also often only very partially clever, but quite stupid in other respects, like to express themselves incomprehensibly, and although they talk a lot about cooperation and teams, they are always fencing out rankings. In other words, they behave pretty normally. At large conventions, therefore, the alpha animals of a discipline, or those who are thought to be alpha animals, are regularly on display. There they show their courtship rituals, with which they try to impress their cronies according to all rules of the art of seduction. *So what ...* a reflection of ordinary life.

Scientists, while supposedly fond of logic and precise talk, de facto probably produce more false than true, and at least as much confusion as clarity. But all this has hardly done any serious harm to science so far; it has always somehow managed to put away the small subjective quirks of its actors. The paradigm disease is different. This seems to be seriously affecting science itself. For if critical thinking is almost switched off for long periods and a factually stable state returns where actually a principled readiness to doubt was always announced, then there is undoubtedly a serious disturbance of the living dynamics of knowledge.

Striving for Truth in Competition with Striving for Power

And there even exists another symptom that – *horribile dictu* – gives even more cause for serious concern about science: the completely changed role that power aspects play in the infested discipline as opposed to the non-infested one. Healthy science is about one thing above all: the search for truth (whatever that may be). Everything else is subordinated to this search for truth, including those subjective sensitivities of individual researchers alluded to above. Science infected by the paradigm virus, however, is immediately recognizable by the fact that it claims a leading role, i.e. that it simply "applies" to the co-infected members of a scientific community. A kind of followerism develops, not unlike that of a cult.

Over long distances, the members of a scientific community "follow" the now entrenched convictions of the infested discipline, similar to the rodents and children of Hamelin following the flute notes of the Pied Piper (an old German folk legend, based on true events), they "obey" them and "follow" their principles dutifully. These convictions "rule" their minds and thus gain, at least temporarily, "power" in a field in which it actually was or should be about quite different categories of goals, namely the uninterrupted effort to find what is true or false. One might think we were dealing with a company of opinionless command receivers, but they are responsible citizens, even more so those who take credit for their specially trained reason. Paradigms rule, even though they are not recognized as true, but merely assumed to be true. In fact, they are simply what has become accepted as the common view, irrespective of their truth status. The truth dimension is abandoned in favor of the power dimension, on the basis of pure belief – this represents the most serious and dangerous symptom of paradigm disease that we know.

But that is still not enough. The peak of perfidy lies in the course of the disease. Paradigm disease, compared to normal viral diseases, usually proceeds quite unusually and dramatically. Whereas the latter quickly or slowly subside after the acute phase of infection and the affected organism is usually completely healthy again afterwards, such a recovery is only very rarely observed in the case of Paradigma. Rather, something occurs which is very unusual in the whole of virology, namely a long-term to permanent succession of Paradigma infestations, only briefly interrupted in each case by bouts of health. Whereas we usually know disease episodes separated by long phases of apparent health, here it is the other way round: a phenomenon that probably explains the widespread misinterpretation that all this is "normal", to which the first describer also succumbed. He has therefore consistently characterized this course as a "paradigm shift" using a term that has since become commonly used and seems to be quite unproblematic. Paradigm infections thus do not normally present themselves as isolated, one-off events, but rather as characteristically sequenced chains of disease phases, between which the comparatively short-lasting health episodes are hardly recognizable.

The innocuously descriptive-sounding name "paradigm shift" does not betray the fact that we are dealing with a disease and, in the case of "shifts", with phases of disease chained together over the long term. It is significant that the paradigm-discoverer Kuhn classifies such shifts as "revolutions," that is, as situations of insubordination and turmoil in which the normal order of things is lost and subversion and anarchy prevail. And this in science! All this shows quite unmistakably the sophistication, even insidiousness, with which the virus operates, which is now in fact globally widespread in all scientific disciplines and is upsetting our categories of normal and successful or disturbed and less successful science. There are even quite a few observers who speak of the paradigm virus or a variety of it infecting not only the sciences but also cultural systems of many or even all kinds, such as art, politics, economics, religion or the so-called Zeitgeist in general. The epidemic would thus already have the characteristics of a pandemic. However, one should exercise conceptual caution here and not overlook the fact that one can once again succumb to deceptions. The temporary mere "getting stuck" of otherwise flexible and changeable ways of thinking is a well-known experience in the mentioned areas since time immemorial, which can also cause massive damage there, but which is mostly accepted as inevitable on all sides. But science is particularly affected by this way of thinking.

Disease of Reason Itself

The reason for this is to be seen in the fact that in science, gaining knowledge in principle and the permanent striving for the goal of truth, in short: rational behaviour, are at the very top of the list of values, while in the other areas mentioned, other, higher-ranking goals prevail, such as profit maximization in the economy, the acquisition of power in politics or the expression of self-realization in art. In science, on the other hand, as an endeavouringly rational activity, truth orientation, flexibility of thought and perpetual critical openness and search for errors should, by definition, have a particularly high value for the system, which – one may regret – cannot necessarily be said for the other areas of activity mentioned.

Therefore, a paradigm infestation and even a succession of paradigm shifts is indeed particularly specific and particularly dangerous for science: these pose existential problems for it. Reason itself seems to fall ill. If, however, rationality itself is to a large extent not healthy but ill, then it should not be difficult to see that science, too, can hardly do justice to its tasks and is in danger of failing as a solution strategy for complex and difficult problems, which are constantly increasing in our world, which is becoming more and more complex. Even the popular theoretical playgrounds on which scientists particularly like to romp around, build beautiful models and argue prettily with each other lose their value the moment the former confuse their toys with reality or have the misfortune to have a crowd of followers choose this as their paradigm. A bad situation! What is to be done?

Difficult Therapy

The question of combating these widespread symptoms of disease therefore presents itself as an extraordinarily serious problem. So far, one must summarize the state of epidemiological research, there is hardly an all-time and sustainable effective antidote known. The disease seems to be practically incurable in most cases, when the short health spurt is replaced by a new, longer infestation.

Mostly, therefore, serious attempts to permanently immunize against the virus or to get out of the vicious chain of paradigm shifts are no longer undertaken at all. Prevention is certainly the best therapy in this case as well. In any case, it seems empirically conspicuous and highly significant that an initial infection with this virus usually results in permanent damage to the affected disciplines. However, it seems to be possible in principle to strive successfully for a significantly changed perception of the problems and objects of a scientific discipline even if one is "actually" convinced that one has found the right or at least a feasible way.

For it would seem that there are two kinds of scientists: on the one hand, those who have such a fear of ignorance and loss of knowledge (and, of course, the associated loss of power) that, under the influence of an infestation of the paradigm virus, they are relieved to see the right path ahead of them, and are therefore quite adamant about methodological rigour, and hold fast to convictions once they have been formed, almost at any cost; but on the other hand, those who take a much looser view. This second species of scientists is the one that ultimately brings with it the better conditions for the progress of knowledge. Their representatives are more open, more receptive to the unexpected and to creative processes of all kinds. Last but not least, they are not such a burden to their scientific community as their more dogmatic colleagues. While the latter do not tire of proclaiming the supposedly found truth and thereby patronize especially the younger scientists who have not yet been established, the others bear it with great equanimity when the younger ones seek and find their own ways.

Our theories of science are, alas, widely theories of scientific soundness: how to ground them logically, methodologically, institutionally, and sociologically is their almost litanically repeated theme. It would help a lot if this solidity fetishism could at least be complemented by an orientation towards scientific creativity. For the reasons mentioned above, this would by no means automatically put an end to the extensive career of the paradigm virus in science, but it would certainly be a substantial contribution to its weakening. Where many scientists today see their task in completely superfluous paradigm reinforcement, it is actually much more valuable to work on paradigm weakening. Only in this way can we combat the fatal tendency to take the mere assumption of truth for more than it is.

More Favorable Prospects of Recovery for Paradigm Shifters Than for Paradigm Confirmers

Many scientists are verbally demanding solidity, but in their own practice they are only more or less consistent in this respect; sometimes things are a bit sloppy. Others belong to the minority of creative minds, but are themselves handicapped in the consistent implementation of their convictions by an unmistakable fear of possibly being considered unsound by their colleagues; they like to demonstrate this by formalizing or mathematizing. This fact now opens up some space for epidemiological therapy. It seems that one is not unconditionally at the mercy of the virus, but that one can take preventive measures above all, but to a certain limited extent also measures accompanying the disease and aftercare, in order to at least limit the mischief it is wreaking. In this context, the change of a paradigm case should not be underestimated. A paradigm changer still has a better prognosis than a paradigm confirmer. For those who, once it has happened, are at least still capable of paradigm shifts, have in principle retained a certain flexibility, which is necessary, if not sufficient, in the further process of combating the virus.

Now, one cannot decide to displace an old paradigm case by a new one, because the decision whether this happens or not is made by the scientific community, not on the basis of logical or methodological criteria, but pragmatically: If the community allows itself to be infected, then it has happened; the state of the concepts that triggered this may be however outrageous. One can be infected precisely – as we have seen – not on the basis of an insight into the truth of a conception, but only on the basis of a belief in it. Science has not only transitions to art (for instance in literary studies) or to politics (for instance in economics), but also an amphibious border zone towards theology. If, however, one is of firm faith and does not want to be re-infected, then a new theory can be no matter how carefully elaborated, no matter how prudently justified, no matter how fully argued: it does not produce a new paradigm disease, it remains with the old one. Scientific communities characterized by this torpor are de facto lost to a return to a state of non-infestation; they seem no longer receptive to any antiviral strategies. The universities in which they are promulgated are no longer sanatoriums (which they should be); they are only nursing homes.

Such communities, however, which have not yet forgotten how to change their paradigms, possess in principle the prerequisite for getting rid of the virus again. However, in fact this never seems to happen completely, but only a part of their members, often only a few, are cured. It is those scientists who have never lost certain doubts about the validity of the prevailing views, although at times they have adhered to them themselves, have always proclaimed the tenets of their discipline with a residue of distrust, and have always remained aware of the loss of coherence that is associated with all specialization. For these people, of course, a form of therapy offers itself which, although it will never be able to gain majority support, can therefore have all the more lasting effect.

These are temporary stays in sanatoriums, retreats in so-called "cross-thinking circles", so to speak, i.e. the communication with other people who are not susceptible to paradigms or who – due to whatever life circumstances – have actually overcome this infection, accompanied by the most favourable external circumstances possible. Mind you: It is not enough to playfully enter into a debate with the representative of another paradigm (the normal scientific case), because this ends like with mating great bustards or capercaillies: There, people do their best to impress and put on an act, and then either one of the roosters goes berserk, changes over to the other as a helper, or leaves the terrain defeated, or both come to some outward arrangement, but don't change their convictions one iota and just wait for the next battle.

With Courage and Lateral Thinking Against Paradigms

Retreats in tried and tested lateral thinking sanatoriums are completely different, because they are not power struggles in disguise, but real playgrounds of free creativity (Feyerabend 1970). They are therefore extremely valuable therapeutic situations. The proximity of scientists who are apparently immune to the infestation of paradigm, or who have at least been able to retain an awareness of the true distribution of health and disease, and who do not participate in the usual reversal of conditions, seems to be of great epidemic hygienic value for the prospects of cure. It is comparable to the copious airing by opening the windows of a room, in which the persons present, for the most part, do not even notice the fug in which they live, unless some one newly coming in calls their attention to it. A "lateral thinker" who is open in his judgement can therefore possibly be of greater importance for progress in a science than all the research projects, however solidly worked out, which – even if they do not notice it – themselves suffer from a paradigm fallacy.

It must be noted, however, that this usually requires a good deal of courage – both on the part of the sanatorium staff and especially on the part of those who are trying to recover there – in order to be willing and able to resolutely withstand the constantly expected new bouts of infection on the part of the almost completely infected environment. Without such a courageous determination to fight against the almost omnipresent disease, the chance of recovery is slim. To those who do, however, the prognosis is quite good. So if one is looking for central categories of a new philosophy of science that is not itself infected by the paradigm virus, one should make sure that civil courage is also one of them. But unfortunately it has rarity value. Paradigm thinking is a typical characteristic of a common understanding of science that has become too involved with political, economic, and bureaucratic categories of power and order to expect most of its actors to mobilize courage anymore (Finke 2018).

It is not a paradigm shift that is needed, but the overcoming of thinking in paradigms (Finke 2005). Some do not believe that this is possible, but it is possible; admittedly, it requires not only consistency and steadfastness, but first and foremost courage. Paradigms are courage killers and precursors of the unjustified megalomania of confusing belief with knowledge and knowledge with certainty, and of basing a claim to power on them. Courage, however, is a more important meta-theoretical category than much of what supposedly belongs to the holy of holies of science, but is in fact only the spawn of a follower ideology infected by the paradigm virus.

Remember how Kant translated his *Sapere aude! Have courage to use your own reason!* So then: Against the paradigms! They are released and may be fought![1]

The Author

Peter Finke (Fig. 6.1) was Professor of Philosophy of Science at the University of Bielefeld from 1982 to 2006 and visiting professor at various universities in Germany and abroad; from 1996 to 1998 he was also Gregory Bateson Professor of Evolutionary Cultural Ecology at the Private University of Witten-Herdecke. In 2004 he was awarded an honorary doctorate by the University of Debrecen in Eastern Hungary

---

[1] First published in "Aufklärung und Kritik" 2/2010, pp. 83–92. We thank the publisher for permission to reprint an abridged and modified form.

Fig. 6.1 Birdwatcher Peter Finke observes the flight of birds in the sky from the horizontal. (Photo: Barbara Bayreuther-Finke)

for his achievements in teaching and research. In 2006, Peter Finke resigned his Bielefeld chair at his own request before reaching the age limit in protest against the encroachment of politics to enforce the so-called Bologna Process, which was abolished shortly afterwards as irrelevant to the Bachelor-Master system. Today he holds guest professorships and is active on various boards on an honorary basis. In recent years he has made a name for himself internationally as a connoisseur and defender of good amateur science.

References

Feyerabend P (1970) Against method. Outline of an anarchistic theory of knowledge, vol 4. Minnesota Studies, Minneapolis

Finke P (2005) Die Ökologie des Wissens. Exkursionen in eine gefährdete Landschaft. Albert, Freiburg

Finke P (2010) Die Paradigmakrankheit und ihre Heilung. Über Diagnose und Therapie einer wissenschaftlichen Epidemie. Aufklärung und Kritik 2:83–92 (Erstveröffentlichung)

Finke P (2018) Lob der Laien. Eine Ermunterung zum Selberforschen. oekom, München

Kuhn TS (1970) The structure of scientific revolutions. UCP, Chicago

7

Scientists, Magicians and Charlatans: How Magic Creates Knowledge

Thomas Fraps

How are magic and science related? A physicist-magician explains the close relationship between wit, information and amazement. In an exciting historical excursion, he illuminates the common roots of the comic, of conjuring and research, overlaps and manipulations, all this also with a view to fake news.

"The most beautiful and profound thing that man can experience is the sense of mystery. It underlies all striving in art and science." Albert Einstein
"From wonder comes joy." Aristotle

At first glance, magic and science seem like irreconcilable opposites: With their illusions, magicians seemingly turn the laws of nature upside down, which science elicits from nature in painstaking research work. The art of magic uses psy-

T. Fraps (✉)
Magician and Moderator, München, Germany
e-mail: thomasfraps@trick17.com

© The Author(s), under exclusive license to Springer-Verlag GmbH, DE, part of Springer Nature 2023
M.-D. Weitze et al. (eds.), *Can Science Be Witty?*,
https://doi.org/10.1007/978-3-662-65753-9_7

chological deceptions and trick-technical methods to create false causal connections that create the illusion of impossible events. Science, on the other hand, seeks true causal relationships, strives for knowledge and creates knowledge based on facts; the illusionary artifacts of magic, on the other hand, are deceptively real fictions.

But opposites attract, as is well known. Especially in the border area between facts and fictions, magic and science touch each other on several levels and enter into an unusual symbiosis, whose history reaches back to the beginnings of the Enlightenment and holds some amusing-scurrile chapters in store. Albert Einstein's opening words hint at the underlying, emotional level: For scientists, the feeling of mystery – usually coupled with the feeling of wonder – is often the starting point, the beginning of the quest for explanations, insight and knowledge. For magicians, the sense of wonder at a playful mystery is the goal of aspiration. In this feeling meet both the astonished audience of the performance of a good magic trick, the methods of which remain hidden, and the scientist who marvels at still unexplained phenomena that nature "shows" him in his experiments. The crucial difference here is that the universe – unlike the magician – does not use deliberate deceptions, or as Einstein put it in a letter to Paul Ehrenfest: "[T]he secrets of nature are hidden by sublimity, but not by cunning" (Einstein 20 June 1923 to Paul Ehrenfest, quoted after Hermann 1996).

For the audience in a lecture hall, for example, this means being able to rely on the fact that the experiments shown and the theories based on or derived from them are concrete facts according to the respective state of knowledge – and not fictions based on trickery or deception. However, it is in the nature of some scientific experiments and theories, especially in physics, that they have a very startling effect on viewers and have highly astonishing consequences that contradict everyday experience. Anyone who has ever witnessed with their own eyes in a live experiment how a stone and a feather (in two vacuum tubes) fall to the ground at the same speed will remember that physics experiments can feel like magic. Other examples of such "miracles of science" include the time dilation effects of special and general relativity or the sometimes absurd-seeming conclusions from quantum mechanics with its wave-particle duality, tunneling effects and quantum entanglements, which Einstein famously called "spooky action at a distance".

The sense of mystery and wonder is here not only at the origin but also at the end of the pursuit of science. It is precisely these astonishing theories and facts that in turn open up an interface for the presentation of science in the context of a magic show that can be funny, informative and astonishing at the same time for a lay and professional audience alike.

"Metamagicum – Miracle Jokes Science"

One example is the program "Metamagicum – Wunder Witze Wissenschaft" ("Metamagicum – Miracles Jokes Science") developed by the author of these lines and the Frankfurt magician Pit Hartling in 2004, which has been performed or is still being performed, among other things, at performances at CERN (world's largest particle physics lab near Geneva), on the occasion of the fifth anniversary of the Wolfsburg Science Center "Phaeno", at (Munich) Science Days, at the annual "Highlights of Physics" events or at high schools and theater festivals and has found a niche in science communication.

Two professional magicians deal humorously with topics from science and technology spiced with a pinch of philosophy. The two protagonists muddle through the four-dimensional space-time at a high level. The program uses, among other things, innovative magic tricks developed by themselves to illustrate the amazing results and paradoxes of relativity theory, quantum mechanics and philosophy and to make them emotionally tangible. The tricks and astonishing experiments are introduced by a factual explanation of the scientific theories and effects, only to be exaggerated in the further course with a wink of the eye and culminate in a magical punch line that leaves the audience laughing and amazed at the same time.

As an example, beaming, based on the quantum mechanical entanglement of two photons, is supposedly transmitted into the macroscopic realm: An spectator's borrowed shoe disappears from a shoebox-like transmitting device and is teleported to an empty receiver box (which looks remarkably like a microwave with an antenna) standing on the other side of the stage, examined by the spectator at the beginning. Another invention is the "Gravitron": a device that, in appearance, could have come from the historical collection of the Deutsches Museum /Munich, but is supposedly capable of "locally altering the Earth's gravitational field". Thus, all of a sudden, not only does a table start to levitate, but – thanks to general relativity and its time dilation in a gravitational field – it also becomes possible to change the flow of time. The gravitron is set to an elevated gravitational field, simulating proximity to the source of a gravitational field, which supposedly makes time run slower on stage. Incidentally, the effect is known from everyday long-distance economy flights even without atomic clocks: If the seats next to you are empty, time runs much faster on an intercontinental flight than in the gravitational field between two massive seat neighbors. As experimental evidence, depending on the gravitron's setting, a borrowed spectator's watch runs faster or slower until time even stops completely and the clock disappears. The gravitron short-circuits, a black hole spontaneously forms, and behind the event horizon, time suddenly runs backwards: empty, dented beverage cans visibly return to their filled, unopened original state, and torn newspapers restore themselves. The assisting spectator disappears from a Polaroid photo initially taken as a souvenir, and the vanished spectator's watch reappears at the end in the sealed peanut can that was given to the spectator as provisions at the very beginning of the time travel.

Elsewhere in the program, the most famous formula in all of physics, $E = mc^2$, is derived from the special theory of relativity. In the strongly abbreviated but mathematically correct derivation, the name of a Munich brewery can be clearly seen on a slate at second glance, which had been "hidden" from everyone's eyes all along in an integral with the letter symbols for the momentum p, the acceleration a, *the* time t, the kinetic energy En and the integral sign itself. This serves as mathematical proof of the widespread anecdote that Einstein, as a youth, had taken a holiday job in his uncle's company and, at the Oktoberfest, had helped to electrify the very beer tent that belonged to the brewery "derived" in the formulas. Fittingly, a postcard of Einstein has been preserved, which he sent to his Swiss friend Konrad Habicht in his annus mirabilis, proudly writing on it that he was "drunk under the table".

These are just a few examples from the "Metamagicum" program, which illustrate how magic art can communicate topics of science on a popular scientific level with wit and – out of amazement – joy. The audience is of course always aware of the illusionary nature of the tricks, apparent technical inventions and experiments presented, even if the accompanying texts – paired with elements of comedy and science cabaret – convey factually correct content. It is always clearly a matter of tricks and deceptions, or as it was called at the time of the Enlightenment, "natural magic" (Brewster 1833).

Between Superstition and Enlightenment

However, if one goes back to the seventeenth and eighteenth centuries, when the Enlightenment thinkers tried to take the mystery out of superstition, witchcraft and the supernatural, while at the same time the boundaries of science and the profession associated with it were not yet sharply defined, so one suspects that travelling magicians, who increasingly included experiments in magnetism, electricity or even chemistry in their programs as "physical amusements", not only aroused amazement in the public, but also uncertainty among the public as to the natural causes of their experiments and thus undermined the idea of enlightenment (Hochadel 2003; Stafford 1998). Especially since both formats, experimental lectures at universities as well as public magic performances in theaters, aimed at spectacular effects in order to win the favour and money of the astonished public. And therein lay the danger, for "in an era when specialization and professionalization [of science] were only in their nascent stages, public displays of experimentation often bore a disconcerting resemblance to magic shows" (Stafford 1998, p. 15). This similarity in aesthetics opened the back door to imposture for magicians and charlatans. A popular representative of these false professors was Jakob Philadelphia (1734–1813). Born in the USA, he performed as an "artist of magic and mathematics" with great success in many European cities. When he gave his first private performances in Göttingen in January 1777, the audience included Georg Friedrich Lichtenberg, who in letters to his friends criticized one trick in particular that was obviously based on magnetism, but which Philadelphia, who was approached, denied (Lichtenberg 1777 [1984]). Lichtenberg took this as an affront and used the whole thing as an occasion for his famous "Avertissement" – a nocturnal placarding of Göttingen with a forged placard allegedly from Philadelphia himself, which promised even more incredible feats than Philadelphia's real placards. For instance, it held out the prospect of switching the weathercocks on the two Göttingen churches "without magnetism only with speed." Philadelphia, exposed by this satire, renounced his already announced public performances and left the city in a hurry. This episode is a pointed example of the difficult relationship between magicians and scientists at the time. Among other things, the Enlightenment scholars attempted to use descriptive definitions to enable a clear classification and to distinguish themselves from the "charlatanry" of a Philadelphia.

> The phisician (physicist): observes the phenomena in nature, seeks to find results from them, which he proves by experiments, seeks to explain the phenomena, but not to deceive his hearers, and is paid for his donated benefit. His ingenuity is admired and he is held in high esteem; he usually stays in one place and receives important state posts.
>
> The sleight of hand: uses the results of the phisician (physicist), makes experiments, but gives no explanation of them, but seeks to deceive his spectators assiduously, and gets paid for this deception. One admires his dexterity, he enjoys little esteem, roams about the country, and receives no state post. (quoted from Hochadel 2000, p. 130)

By the way, from the personal experience of the author, a professional magician and graduate "phisician", this definition can still be largely confirmed today. And so it is a beautiful irony of history when exactly one hundred years later this initially so difficult relationship takes a surprising turn and the magicians take over the tasks of enlightenment, as some representatives of science fail and involuntarily abet the supernatural, which in the form of spiritualistic séances and sessions of self-proclaimed mediums finds its way throughout Europe.

The Zöllner-Slade Controversy

The US-American medium Henry Slade (1836–1905), for example, succeeded with his séances in London in 1876 in winning over, among others, Sir William Crookes, physicist and discoverer of thallium, the mathematician Lord Rayleigh as well as Alfred Russel Wallace, the co-discoverer of the theory of evolution, as convinced advocates (much to the horror of Charles Darwin). Above all, Sir William Crookes, after his experimental tests, denied any kind of fraud to the medium Henry Slade, thus granting him scientific legitimacy. It even came to an indictment of Slade and a famous court case, in which the demonstration of spiritualistic effects by the magician Sir Neville Maskelyne ensured that Slade was convicted. However, the latter was able to flee to Germany before the sentence came into effect, and there he met probably his greatest follower, Karl Friedrich Zöllner (1834–1882), the first German professor of astrophysics, in Leipzig. A decisive factor for his belief in spiritualism was that Zöllner had already been working for some time on a theory of the fourth dimension of space – a hypothesis that Riemann, Helmholtz and Klein, among others, had also been working on. Zöllner saw Henry Slade as a kind of measuring device, a mediumistic instrument that had access to the fourth dimension (Staubermann 2001). Even though many scientists of the time were very sceptical about spiritualism, Zöllner was not alone; his physicist colleagues Wilhelm Weber and Theodor Fechner were also present at the experiments with Slade and were themselves convinced followers who considered Slade's spiritualistic abilities to be real and deception to be absurd.

> The statement of physical facts, however, falls within the domain of the physicist; and when men of such outstanding importance as Wilhelm Weber, Th. Fechner, and others, openly advocate the reality of such facts after thorough experimental examination, it is obviously nothing but an act of modern presumption on the part of the unscientific public when the latter indulges in accepting ridiculous conjectures about the possibility of a deception as fact without further ado, and thereby denies those men the ability to make exact observations. (Zöllner 2008, p. 79)

The psychologist and philosopher Wilhelm Wundt, however, doubted precisely this claim of Zöllner. He insisted that scientists were only authorities in their own field and that the séances of a Henry Slade were outside their sphere of experience and thus their authority as scientists. He himself did not trust Slade's alleged abilities and felt that none of his phenomena "went beyond the performance of a good sleight of hand ... [and] it might not have been altogether improper to have taken a closer look at the performances of a dexterous conjurer" (Wundt 1879, p. 401). The dexterous conjurer in this case was Carl Willmann, one of the best-known magic-device dealers of the time. In his 1886 book Modern Miracles, he analyzes and explains the tricky maneuvers employed by Slade and other mediums, for "...

numerous debunkings furnish proof that fraud plays a prominent part in spiritualistic sessions …" – and so, as a magician versed in deceiving the senses, he "could not help smiling at the credulity of the gentlemen scholars" (Willmann 1886, p. 154 f.). Here, once again, the feeling of mystery, the amazement at the inexplicable – as in the time of the Enlightenment – proves to be the Achilles' heel of reason and, at the same time, an emotional back door for swindlers and charlatans. Whereas in the Enlightenment it was only the naïve spectators among the people who, marvelling, often did not know how to distinguish between genuine experiments, tricky magic tricks and seemingly supernatural powers, in the case of spiritualist mediums it is now even the scientists themselves who succumb to the tricky deceptions.

The Geller Controversy

From today's perspective, the Zöllner-Slade controversy is not without a certain unintentional humor, even if one takes into account the contemporary historical-religious context in which the séances took place. But Wilhelm Wundt was to be proved right. Almost a hundred years later, in the mid-1970s, the story of misconceived authority and scientific hubris regarding paranormal phenomena was repeated once again. Several experienced scientists from around the world confirmed to the Israeli medium Uri Geller (*1946) that he did indeed possess telepathic and telekinetic abilities. The conducted experiments showed similar negligent errors as hundred years before and even led to a publication in the highly respected journal "Nature" in 1974 (Targ and Puthoff 1974). The authors Russel Targ and Harold E. Puthoff were laser physicists who conducted their experiments with Geller at the Stanford Research Institute, a private research institute spun off from Stanford University in 1970. Geller duplicated drawings sealed in an envelope, including guessing the top number of a cube protected in a small metal container eight times in a row. The success of these laboratory experiments was presented as (scientific) proof of Geller's telepathic abilities.

However, as the magician and declared Geller opponent James Randi describes in his book "The Truth About Uri Geller", Geller – contrary to the description in the "Nature" article – was allowed to touch and handle the box with the cube himself after it had been shaken by the experimenter – a small but crucial detail for magicians. This information, together with the fact that the cube was protected in the container only by a removable lid and not by a lid with a lock, allows the explanation, obvious to a magician, that Geller, by means of dexterity, was able to lift the lid surreptitiously and to catch a brief glimpse of the number on top of the cube through the slit (Randi 1982).

The editors of "Nature" still remarked in the preface to the article by Puthoff and Targ (1974) that they were convinced after consultation with the authors that Geller's effects "cannot be explained by standard magic tricks". Standard manipulative methods used by Geller were certainly not, but they were still trick methods. The theoretical physicist David Bohm and his former colleague Jack Sarfatti had also witnessed a demonstration by Geller at Birkbeck College in London in July 1974. Both were convinced of Geller's abilities after thorough tests. The latter had, among other things, bent a borrowed key from Bohm and caused a Geiger counter to deflect several times, so that Sarfatti published a press release with the following conclusion: "My personal judgement as a doctor of physics is that Geller demonstrated true psychoenergetic abilities at Birkbeck under relatively well-controlled and repeatable experimental conditions" (Sarfatti 1974, p. 46). The physicist thus falls into the same psychological trap that had probably doomed Zöllner and his colleagues a hundred years earlier: namely, believing that as a "doctor of physics" one was apparently immune to simple deceptions. The amateur magician and science journalist Martin Gardner reports:

> When Sarfatti was asked if anyone had searched Geller for a radioactive beta source, he was told by Sarfatti that no one had thought of such a possibility and that it was a brilliant idea. Magicians find this answer merely comical. (Gardner 1983, p. 73)

Thus, from today's perspective, one looks back on this Geller controversy not only with a frown, but also with a smirk. The perceived superiority, however, which creeps up on you while reading, is due to the temporal perspective and is quickly put into perspective when you consider that similar cases still occur today – but with a different coloration. Now, however, it is no longer ghosts or supernatural forces that are cited as false explanations, on the contrary: some of the mentalists, the self-proclaimed mind readers of the present, explain their – merely feigned – amazing abilities themselves with selectively chosen set pieces of science: from NLP to cognitive psychology, hypnosis and epigenetics to the reading of body language signals and studies on mirror neurons.

Between Fact and Fiction

We are dealing here with a double deception appropriate to the post-factual age. For when asked, "How do you do it?" the answer does not invoke supernatural forces as it still does in the case of Slade or Geller, but pseudo-scientific bogus explanations that feel emotionally plausible because they dissolve the cognitive dissonance of wonder into scientific pleasantness. The enlightened person in particular is apparently susceptible to bogus explanations given under the guise of science. And so it happens that science journalists on public television (ZDF, Schmidt 2015) shove a mentalist into the brain scanner at prime time in order to examine his empathy and "special empathic ability" in the laboratory and explain it in front of an audience of millions with "clearly increased activity of the mirror neurons". The fact that the phenomena demonstrated in the show by the mentalist are only stagings based on trick techniques is not mentioned.

Another example is the science editorial team of the show "Mich täuscht keiner!" (No one fools me!) (ZDF 2017), which was fooled by a mentalist who, during a live demonstration in the studio, claimed to be able to recognize lies based on reading the body language of prominent candidates and to be able to assign drawings made by the candidates to their respective authors. Not a word about the actual trick method, that the white drawing boxes were marked with pencil dots and handed out by himself to the four candidates in a certain order at the beginning. The presenter of the show did not even ask the question whether it was deception or not, because according to his own statements (when asked by the author) the responsible editors had no idea at all that it could be a trick. The whole thing takes on a particularly ironic note, since the central concept of the show was precisely to reveal and explain the many facets of deception: from optical illusions and animal camouflage to shell games, con artists and tricksters.

And then there are the numerous non-fiction and advice books by some mentalists, which have been thrown onto the book market for years, some of which have become bestsellers, thus spreading the scientific bogus explanations as fake news (Jan Becker, Thorsten Havener, Tobias Heinemann, Norman Graeter etc.). As a result, these are not only believed by spectators, but also presented as facts in newspapers: "It is not a matter of his demonstrations being tricks and illusions that mislead his spectators in order to amaze him …", writes e.g. the Süddeutsche Zeitung in its review of a mentalist performance (SZ 2013). The boundary between facts and fictions, between science, pseudoscience and magic tricks is thus blurred for entertainment and marketing purposes – and at the expense of science.

Conclusion

And so we have come full circle: we are back to Philadelphia & Co., who used scientific phenomena to present amazing things to their paying viewers, and were not so careful about the truth. Today it is no longer physics and chemistry, but psychology and neuroscience that are suitable as bogus explanations. But

> … just because a good magician demonstrates something extraordinary, you shouldn't jump to the conclusion that it's a real phenomenon; you need a lot more evidence for that. But it's fun to figure out the trick, and the only way to figure it out is to be completely sure it's a trick, and not be willing to believe it isn't, because then you slip too easily.

as Richard P. Feynman wrote about an encounter with Uri Geller (Feynman 1989, p. 49 f.).

The emotions of mystery and wonder not only arouse curiosity, but also briefly suspend our cognitive-rational coordinate system. They therefore not only serve as creative driving forces in science and art, but also prove to be the Achilles' heel of the Enlightenment in the anecdotes described – and this continues to the present day. Even scientists slip too easily on (account of) these emotions.

The invisibility of causes, which makes us wonder, and the invisible boundary between facts and fictions complement each other in this case in an unfortunate way. Picasso's observation that art is a lie that makes us see the truth applies to the art of magic insofar as it makes the existence of the limits of our perception playfully visible. In the case of past charlatans and present-day con artists, however, the artistic nature of the lie is absent, for the true nature of the causes is deliberately left in the invisible. The joke is that in these cases it is the magician, of all people, who can lend a hand to the scientists in their search for truth, to make visible the difference between sublimity and trickery, between facts and fictions.

The Author

Thomas Fraps (Fig. 7.1) is a professional magician and adult. For the first 27 years of his life he often wrestled with reality, but in the end he won! Since then, he has roamed the world as a magician, playfully turning upside down the laws of nature he previously learned as a graduate physicist. Thanks to his very special theory of reality and amazingly entertaining magic, he brings the beautiful feeling of amazement to his audience's memory and creates magical moments that entertain in the best sense, whether at a company party, a private party or in the theatre.

Especially in his famous role of the "False Expert" (http://www.thomasfraps.com/derfalscheexperte.html), Thomas Fraps, as an amazing comedy speaker, provides "frapp(s)ierende" (German wordplay for striking) moments at

Fig. 7.1 The magician Thomas Fraps multiplies himself. (Photo: Gerald F. Huber)

international conferences, symposia and specialist meetings of all kinds. Whether at a supercomputer conference in San Diego, a neuroscience congress in Paris or at the ceremonial opening of a sewage sludge incineration plant in Schongau – no audience is safe from the False Expert.

References

Brewster D (1833) Briefe über die natürliche Magie an Sir Walter Scott. Enslin, Berlin

Feynman R (1989) Mr. Papf's perpetual-motion machine. Skeptical Inquirer 14(1):49

Gardner M (1983) Kabarett der Täuschungen – unter dem Deckmantel der Wissenschaft. Ullstein, Berlin

Hermann A (1996) Einstein der Weltweise. Piper, München

Hochadel O (2000) Nur Taschenspieler und Scharlatane – Wissenschaftliche Schausteller in der deutschen Aufklärung. In: Hochadel O, Kocher U (Hrsg) Lügen und Betrügen. Böhlau, Köln, S 113–131

Hochadel O (2003) Öffentliche Wissenschaft. Elektrizität in der deutschen Aufklärung, Wien, Wallstein

Lichtenberg G (1984) Brief an Georg Heinrich Hollenberg vom 9.1.1777, sowie Briefe an Andreas Schernhagen. In: Joost U (Hrsg) Der Briefwechsel zwischen Johann Christian Dieterich und Ludwig Christian Lichtenberg. Vandenhoeck & Ruprecht, Göttingen (Erstveröffentlichung 1777)

Randi J (1982) The Truth about Uri Geller. Prometheus Books, Amherst

Sarfatti J (1974) Off the beat: Geller performs for physicists. Sci News 106(3):46 (20.07.1974)

Stafford B (1998) Kunstvolle Wissenschaft. Philo Fine Arts, Hamburg

Staubermann KB (2001) Tying the Knot: Skill Judgement and Authority in the 1870s Leipzig Spiritistic Experiments. British J Hist Sci 34(1):67–79

SZ (2013) Süddeutsche Zeitung (Landkreisausgabe Fürstenfeld-bruck), 19.11.2013, S R11

Targ R, Puthoff HE (1974) Information transmission under conditions of sensory shielding. Nature 251:602–607

Willmann C (1886) Moderne Wunder. Otto Spamer, Leipzig

Wundt W (1879) Der Spiritismus. Engelmann, Leipzig

ZDF (2017) Mich täuscht keiner! https://presseportal.zdf.de/pm/mich-taeuscht-keiner/. Accessed: 20 Aug 2019

ZDF, Schmidt A (2015) Terra X. Die Geistesgiganten. https://www.zdf.de/dokumentation/terra-x/die-geistesgiganten-100.html. Accessed: 20 Aug 2019

Zöllner F (2008) Vierte Dimension und Okkultismus. Bohmeier, Leipzig

8

Searching for Humor in the Deutsches Museum: An Exploration

Wolfgang Chr. Goede

Jürgen Teichmann is Germany's longest-serving museum educator. With students from the TU Munich TUM, he set out to find humor in the Deutsches Museum. Mission Impossible? No, humor germs can indeed be found. But how can they be made to flourish there and at other sites of the natural sciences?

Jürgen Teichmann, a physicist with a post-doctoral qualification, has spent almost his entire professional life on the Museum Island in Munich. He built up the astronomy department, was later museum director, and always travelled a lot around international museum worlds on the trail of fresh ideas to exhibit scientific things in an even more exciting way. For twelve years now, in retirement, he still cycles

W. C. Goede (✉)
Science Facilitation, Munich, Germany

83

daily to work in the historic building in the middle of the Isar river. In the TUM seminar "Can science be funny? Criticism and Humor in Science Communication," held at the museum, he is initially a curious listener. Then comes his act. He guides the students through the astronomy department "in search of humor", according to the joint work assignment.

Astrology or Astronomy?

"Humor was not yet a topic in our time," explains the 77-year-old on the fifth floor of the museum tower, which leads to the "Reich zum Himmel" (Kingdom to Heaven) and the observatory. At the entrance is the brittle word "Astronomie/Astronomy," which prompts Teichmann to play around with it a bit. "Astronomy, but beware: NOT astrology," it might read there to lighten things up, he suggests. Alluding to the fact that the two terms are constantly confused: serious astrophysics and the derivation of fate from planetary constellations – astrology, for many logicians pure reading the leaves.

An important exhibit on the tour are Fraunhofer's lines. It was the Munich optician who catapulted celestial science a giant step forward with his discovery in 1814. When starlight is dispersed with a prism, lines appear that can be used to characterize the suns in the night sky according to temperature and gas chemistry. This results in a categorization of O, B, A, F, G, K, M with decreasing surface temperatures, from which the Anglo-Saxons derived a mnemonic: "**O**h **B**e **A** **F**ine **G**irl **K**iss **M**e".

Later, as more and more women entered science, the saying struck the research world as chauvinistic, so the girl was expanded to "Girl/Guy". The information text in the Deutsches Museum still contains the original version, like all writings relatively small and difficult to read. From this, from social upheavals and contemporary historical adaptations, a narrative could be developed, perhaps even a lively quiz, as Teichmann notes in other exhibits, to increase interaction with visitors and the entertainment character of exhibits. Bavarians, for example, have translated the mnemonic into the regional idiom: "**O**hne **B**ier **A**us'm **F**ass **G**ibt's **K**oa **M**aß" (Without beer from the barrel there's no stein) – very humorous, there's certainly more to it!

Black Humor

Yes, women and science could make up a separate section, even a museum of its own, that would focus on the struggle of the female sex against a traditional and power-conscious male domain. Here, wit and humor could join forces with their twin siblings, namely satire and parody. Take neutron stars, for example: in 1967 the first celestial body of this kind was discovered, a star shortly before its end, collapsed into an extremely massive small body that emits strong pulses of light at high frequency like a beacon.

The inscription in the German Museum bears witness to the joke of the scientists at the time, who christened the neutron star LGM-1, "Little Green Man" – in analogy to an alien who drew attention to himself or perhaps even sent SOS signals. There is indeed human tragedy behind this. PSR B1919+21, the official acronym of this pulsar, was discovered by PhD student Jocelyn Bell, but it was her thesis (male) advisor who was awarded the Nobel Prize for it in 1974. A piece of black humor in the history of science, which is not exactly poor in this respect, in which so many research efforts and dreams were shattered. A gold mine for courageous cabaret artists.

Museum Loriot and Tegtmeier Formula

The astronomy exhibition is still one of the best-visited departments in the Deutsches Museum. Visitors from all over the world, especially young ones, scurry around. Nevertheless, Teichmann is well aware of the exhibition's shortcomings in the light of modern museum education. "There's far too much text," he says. "There should be more games, especially for the smartphone generation, which is used to swiping." Long nerdy explanations should be broken up with more pictures and comics, interspersed with a video featuring a cabaret artist who reveals something funny about the exhibit and the story of its discovery and application.

A museum Loriot (famous German cabaret artists), what would he do to attract crowds to the exhibitions! "Because everything is much easier to absorb and retain when you're laughing merrily," Teichmann knows. Further suggestions come from the TUM students. "Two clowns would be enough," suggests a fellow student, obviously still inspired by the performance of a humor professor in this seminar series, "one who lectures in a professorial manner and another who constantly interjects" – according to the motto: There are no stupid questions, only stupid answers.

For Teichmann, "witty counterpointing" would be a new benchmark for museum communication, as he explains during the tour with the TUM students. He refers to the Ruhr comedian Jürgen von Manger, alias "Tegtmeier", who set up a scientific-cabaret memorial to Otto von Guericke and the proof of air pressure with the help of the Magdeburg hemispheres (Fig. 8.1) with his natural wit. Guericke had put the hemispheres together, pumped out some of the air and thus created a vacuum inside. He then had eight horses

Fig. 8.1 The Magdeburg hemispheres: They, too, can become a lesson in scientific-cabaret humor. (Photo: Deutsches Museum, Munich)

harnessed to the left and right to pull the hemispheres apart against the external pressure.

"What was Guericke's achievement?" asked Tegtmeier rhetorically, and immediately answered himself, "Finding sixteen horses so weak that they couldn't tear the hemispheres apart." That's how science humor walks along, plain and simple. The Tegtmeier formula of turning a content upside down and providing it with an unexpected negation could be applied to other historical experiments with audience appeal and ease.

Moon Diet

The Pluto attribute "the degraded planet" is also ingenious, the students rejoice. With its downgrading to a dwarf planet and expulsion from the majestic solar system, the celestial body made headlines twelve years ago and continues to divide celestial scientists to this day. Catchy anecdotes of this kind, according to the unanimous opinion of the TUM junior academics, are lacking and should be increasingly curated by museums. Artistic resources are especially necessary in highly complex disciplines such as biochemistry, they say.

The cosmic scales are a big hit on the TUM excursion through the astronomy department, still sparkling despite the patina of decades. It tells visitors how much they weigh on the various planets in the solar system. The greater the mass, the heavier the live weight, and vice versa. On the massive Jupiter, a person weighing 75 kg would weigh 180 kg, on the Earth's moon only an almost featherweight 15 kg. A student immediately pulls a marketing idea for low-calorie food out of the hat – "the moon diet". Laughter, applause, super atmosphere: it's so easy to humorously lighten up the heaviest gravitational bodies.

The Author

Wolfgang Chr. Goede is co-editor and has already been introduced in the introductory chapter.

9

From Big Bang to Big Van

Helena González Burón
and Oriol Marimon Garrido

Big Van Ciencia is a Spanish science cabaret – since 2013, with 642 performances in over 20 countries around the world and a quarter of a million participants, from prisoners to ministers. Here, two members of the troupe tell how funny science goes, reveal recipes and how it can make STEM subjects really exciting.

Some say that humans are the only animals that can laugh. Well, the truth of this assumption varies, just think of the cute kittens on YouTube. What we can say with absolute certainty, however, is that humans are the only

H. G. Burón (✉) • O. M. Garrido
Big Van Science Cabaret, Barcelona, Spain
e-mail: helena.gonzales@bigvanciencia.com;
info@bigvanciencia.com

© The Author(s), under exclusive license to Springer-Verlag GmbH, DE, part of Springer Nature 2023
M.-D. Weitze et al. (eds.), *Can Science Be Witty?*,
https://doi.org/10.1007/978-3-662-65753-9_9

creatures that have mastered science and can use it to communicate with their fellow species.

How does communication work in general and in particular? We had always guessed it: with humor. Humor connects people, makes the learning process entertaining, tears down the barriers of power. Humor generates laughter, and laughter is contagious, which is why babies smile back when smiled at. We have all learned that positive emotions are more contagious than negative ones, and those who display positive emotions are said to be less prone to illness.

From all this follows the equation: laughter is healthy, strengthens health and is a social glue. In this respect, it seems logical that laughter can be used to spread science among people and infect them with an affinity, even passion, for it. With this we would have formulated our hypothesis, which would have to be proven in the following.

First of all, any approach to scientific truth, however self-evident, requires experimentation. That's what Big Van Ciencia, a troupe of mostly PhD scientists with a great love of art, takes care of. On stage they create a kind of stand-up comedy with which they entertain their audience, make them laugh and educate them scientifically in the process.

Jesters of Science

Many of our guests think that we are the court jesters and Eulenspiegel of science. Quite rightly so. With wit and humour, we have been able to address the most diverse social strata with science. From young people who had no idea what to do with it before, to highly paid professors who knew much more than we did about the questions raised. Nevertheless, everyone laughed heartily at our jokes about biology, chemistry or mathematics. But also about delicate topics such as the precarious employment conditions of many researchers, their miserable pay, their far too uncritical attitude of mind, the still almost insurmountable gap between the sexes in science ...

In one sentence: humor gives us the fool's licence that was once only granted to court jesters at princely courts, namely to question science and to rethink its role.

Club Appearance for Science

Big Van Ciencia was born in 2013, when its twelve founders met at the semi-finals of FameLab España. Every year, scientists from more than 30 countries compete to give the best presentation. In no more than three minutes, they have to explain a research question to an audience with no scientific background, usually in a very matter-of-fact way. We wondered how this could be spruced up, for example with humor. That's how we came up with our original name, "The Big Van Theory," which we promptly renamed "Big Van Ciencia" (ciencia is Spanish for science) because we didn't want to steal the show from a similarly titled TV series.

Our first gig in brand new t-shirts with our logo was at the legendary "Frikoño", a creative freak festival in Logroño, northern Spain. The visitors were so enthusiastic about us that we immediately decided to go on tour. From the province we went to the big city Valencia, where a disco with the nice name Opal-Club had booked us. We arrived, our hosts showed us the equipment and said, "... here's where you can plug in your guitars ..." Big Van, they had assumed, would be a band. "What, you're scientists and all you need is a mic?" Whew.

We mastered the cliff. Instead of Metallic Sound, we were beating the disco-goers over the head with particle physics and maths, without anaesthetics or anaesthesia. They not only kept still, but were incredibly inquisitive and asked us a lot of questions. "Metallic Science" was an option – and after this performance we were hell-bent on spreading our mission all over the world.

The twelve nameless scientists grew together to form an extended family of currently 25 members from all corners of Spain. They come from all scientific disciplines, now including astrophysics and information technology. As you can see, we are decidedly interdisciplinary, not to say inclusive. To date, we have held 642 events, as is scientifically accurately recorded in Excel tables.

From New Zealand to Turkey

Among them, performances in Mexico, such as at the International Theatre Festival in Cervantino or at the world-famous International Book Fair in Guadalajara; in the Cervantes Cultural Institutes of our country in Naples, Oran, Casablanca (Italy, Algeria, Morocco), in the Liberarte in Buenos Aires, in the UNESCO headquarters in Paris in front of festively dressed delegates from all over the world, in the Borrás, Capitol and Teatreneu in Barcelona, in the Alcazar in Madrid, in the Talía in Valencia. We also kept a record of audience numbers. A quarter of a million visitors in over 20 countries have seen us, from New Zealand to Colombia and Turkey.

The social background and educational background of our clientele is just as diverse as the geographical mix. In conventional venues such as museums, theatres, festivals, we gather academics, the middle class, the educated middle class around us. But we also perform on streets, markets, even in prisons. Or in ministries. And everyone, from state secretaries to residents of slums, laughed at our joke science. In the end, we even ventured into educational institutions and did our thing there. And there, against all expectations, we were celebrated as superheroes.

What we feared: Scientists, let loose on a pack of bored bums waiting for the first opportunity to boo and ridicule the performers. It turned out differently: they didn't laugh at us, they laughed with us. Sure, science can be funny, and how!

Enthusiasm everywhere: We had to pose for selfies with the students, wrestle with a whole avalanche of curious questions, such as what happens when you get into the maw of a black hole, why your intestines growl when you're hungry. What one actually did as a scientist – although hardly anyone wanted to believe that we were such scientists ourselves. Counter-question: "What does a scientist look like?" The young people had stereotypical ideas based on the types drawn by Hollywood and the media: from the lovable "Doc" from "Back to the Future" to the nerdy Sheldon Cooper in "Big Bang Theory". It grabbed us in our professional honor and self-esteem. And besides, we asked ourselves: Where are the barriers that keep so many students from pursuing a scientific path?

Critical Science

This concern gave rise to PERFORM (www.perform-research.eu), an educational study with young people as the target group, funded by the European Commission. Our goal is to help the study of STEM subjects (Science, Technology, Engineering, Mathematics) get back on its feet. To do this, we use a cabaret twist à la Big Van Ciencia. After all, stagecraft is not only a way to enrich scientific content, but also to incorporate the humane demands of science, its values and ethics (as required by the European Union's Framework Programme on "Responsible Research and Innovation").

For this, you have to know your audience very well, the fears of the adolescents, their doubts, but also real existing barriers that stand in the way of a STEM career. This cannot be mastered with questionnaires, but with lively workshops full of creativity and improvisation, humor and reflection. PERFORM incorporates theatrics and action to show what science is, as well as critical thinking about the methods of science: what social challenges and constraints does science face today? How gender-sensitive and fair is it? What hurdles does the job market present? And everything else that moves our participants.

We develop the questions and answers in a participant-centered and participatory way. Art and humor are our tools, which helps enormously to bridge the gap between young people and science. We have taken PERFORM to schools in Spain, England and France. Crossing borders and cultures, we want to show that humor can be used to break down distance and authority – mind you: we are not questioning respect here, but submission and subservience in the traditional education system! Play, laughter and improvisation create a trusting framework that allows students to speak their minds openly.

The results are conclusive. Humor and art ignite a desire for knowledge and science in young people who previously had nothing to do with it, and ignite it in those who were already interested. The participants became – we hope – more open-minded and critical, not only about science topics, but overall. Humor, play, art are democracy-building, inclusive, universal. And almost nowhere in the world did we encounter humorless people.

Create Cross Connections

Science and humor in tandem – that works, as our experiences and examples show, which proves the hypothesis from the introduction. However, many are not yet satisfied with this. Academics in particular want to know whether we follow a clearly defined methodology in our scripts and in our performances. Creativity is difficult to put into rules, but we would like to share a few recipes.

Our mantra: Understandable communication of science must be the main goal. However, this must be designed in such a way that it engages everyone, the audience and us as well. This requires a story as a frame, which we create through associations, i.e. mental connections: Connections between the material we teach and situations that are atypical, abnormal and curious for the participants. This comes across as surprising and funny when, for example, we equate the logic in mathematics with the logic of a partnership, or underpin bacteria and our war against them with "Star Wars" and galactic campaigns of conquest. The crazier the cross-reference, the more exciting, lasting and memorable it is for the audience – and the more grateful they are afterwards. It reads easy, but it isn't in the implementation, especially when it has to be spontaneous and often ad hoc in the play sequences.

Exercise Creativity

Creativity is the magic word. In other words, the ability to think around corners – "out of the box", as the Anglo-Saxons like to say – to associate freely and to create new, fresh contexts pictorially. To avoid any misunderstandings:

no work of art is produced in the process! Rather, the aim is to develop fluid and original ideas, for example for dealing with the burdens of our modern civilization such as global warming, car traffic and congestion or plastic waste, and to embed these in a humorous text.

The popular belief that a creative person mainly uses their right brain while the rational person relies on the left is a myth, because: Creativity is innate in all of us and is also heavily used by the "rational". So the question is not whether we are more or less creative by nature, but whether we are more or less trained in being creative. According to recent scientific understanding, creativity, ingenuity and inventiveness result from the number of neural connections between different parts of the brain. The more of them there are, the more creative we are – and this can be practiced.

This leads to the core question of how we train the creativity that is so fundamental to everything in science education. The answer, as simple as it is straightforward, is to build neural bridges in minds and weld cross-connections! Why don't we start freely from the bottom of our lungs by revitalizing the STEM subjects, which are unpopular with students, however with the arts we turn STEM into STEAM (STEM subjects supplemented by the Arts). Could we, in a great educational reform project, incorporate theatre and literature into the teaching of biology, and vice versa physics and mathematics into the teaching of the arts; and ultimately place stagecraft, drama, improvisation and humor in all these subjects?

Our show juggles with these elements and recipes. The narrative leads us through the science theme, makes the abstract vivid, brings it to life and gives the piece the stamp of "human". To do this, we tell personal stories that have happened to us while doing research. In doing so, we also draw from a reservoir that is well filled by curious audience questions such as: What oddities do you encounter in your lab work? We incorporate all of this into our monologues and dialogues with the audience, and we focus them on one statement: science is at the service of society, providing answers to societal challenges.

Professional performers of stand-up comedy rely on clichés to get their audiences to laugh. This is tempting, but if we simply copied it, we would only be serving the negative image of research and scientists, and thus defeating our purpose of making people curious about research. How do we avoid the black ice? With creativity!

We create completely contradictory and absurd situations, brush people against the grain. The more over-the-top, the more effective. That's humor, that's what generates laughter, that's our craft. Here's an example: In our sketches, we often present ourselves as lonesome wolves and cranky loners, hermits in laboratories, surrounded by test tubes and equations, on the trail of strange things … but then, bingo, the twist. Which the audience immediately honors: Scientists who dare to take the stage, flesh-and-blood to touch, who make the auditorium smile, burst out laughing, not with clichés, but with stories the audience can identify with, who end up targeting the problems of their lives one on one. Fanfare.

Here Come the Science Bards

Time to take stock: after more than five years in science entertainment, we know that there are people everywhere in droves who not only want to be informed, but also excited, not just taught, but downright hooked. We are happy to pass on the lessons learned on the stages and in the classrooms of the world to all scientists who want to inoculate citizens with a passion for the profession, but who also want to contribute to a more critical attitude. That is why we have designed continuing education courses for our colleagues. They are called "Telling Science" and take place at universities and colleges throughout Spain, Europe and Latin America. In addition, together with UNESCO, we have created the "Science Slam-LAC" project, which trains the oral communication of science. This gave rise to a group of stand-up scientists in Uruguay who call themselves "bards". Thus, the historic Celtic poets and singers are experiencing their resurrection as scientists.

The Authors

Helena González-Burón (Fig. 9.1, far right), PhD in Biomedicine (2014), has always very intimately connected her scientific studies with theatre activities. In 2013, she

Fig. 9.1 Oriol Marimon-Garrido (second from left) and Helena González-Burón (far right) in action: they perform a pantomime with a birthday audience and teach how to make something burn with a flint. (Photo: Wolfgang Goede)

co-founded Big Van Science, a non-profit organization that brings science to the people with theatrical elements, humor and storytelling. She describes her profession as a science comedian, science teacher, trainer in public communication of science and science writer. She has organized numerous science education and training projects in Europe, Africa and Latin America.

Oriol Marimon-Garrido (Fig. 9.1, second from left), PhD in Chemistry (2013), is also co-founder of Big Van Science. He specializes in science stand-up shows for young adults and clown shows for children and families. He coordinated the EU project "PERFORM". It aims to find new methods for school pedagogy that use art to arouse and encourage scientific interest in school children. He is currently developing training courses in oral science communication for teachers as well as researchers at universities around the world.

http://www.bigvanciencia.com/

10

When a Dalmatian Comes to the Cash Register

Eckart von Hirschhausen

Eckart von Hirschhausen is the inventor of medical cabaret. The medical doctor is now one of the most prominent humorists nationwide: on television and on stage, in bestsellers and with his own health magazine. Here he reports on the first scientific studies to measure humor. He explains what humor is and how it triggers laughter; he reveals important rules that will also make your jokes work. But above all, the doctor argues: Comedy, wit and cheerfulness are the best medicine against illness.

A man walks through the streets clapping. Another asks, "What are you doing?" Answer: "I am driving away the elephants." Inquiry, "There are no elephants here." Answer: "As you see!"

E. von Hirschhausen (✉)
HERBERT Management, Frankfurt a. M, Germany
e-mail: hirschhausen@hirschhausen.com

© The Author(s), under exclusive license to Springer-Verlag GmbH, DE, part of Springer Nature 2023
M.-D. Weitze et al. (eds.), *Can Science Be Witty?*,
https://doi.org/10.1007/978-3-662-65753-9_10

There is no shorter way to summarize the human tendency to false causality and hubris. Do messages become unserious just because they are understood? Like Paul Watzlawick, as a doctor and science journalist I am convinced of the healing and enlightening effect of humor. I've made a living from it for 25 years. And I've been taken seriously for the last five years, which pleases me greatly. For a long time I was pretty much alone as the inventor of medical cabaret. How do you recognize pioneers? By the arrows in their backs. For decades it was a tough struggle to decide which pigeonhole I belonged in. For the science editors I was too funny, for the entertainment editors too serious. And that's a very German problem, because our brain doesn't distinguish between serious and funny, serious and light music, ARD and ZDF (the two public German TV channels). It only distinguishes between boring and interesting. And for something to be "strange", it has to come across as a bit strange.

I discovered my love of communicating medical facts in a humorous way while I was still studying in Heidelberg. At that time, I was supposed to write an article about venous diseases for the local Rhein-Neckar-Zeitung and wanted to make it vivid: When the inner blood vessels are clogged, the blood takes the way through outer vessels. You can see these congested vessels. So I wrote: Imagine the tunnel through the mountain is closed, then automatically more people drive along the shore road. That's how varicose veins develop. That, however, wasn't medical enough for the editors at the time and the comparison was removed. Today, 25 years further on, I have my own magazine ("Dr. Eckart v. Hirschhausens Stern Gesund Leben"), in which I am allowed to write as I like. But it was a long way.

No one suspected where this would lead when I was given my first magic box at the age of eight and started collecting jokes. What I learned about deception and comedy from scratch still shapes my thinking today. So do all the years I spent studying medicine and learning about the world of health. The entertainer in me is older than the medical professional. Many say, how can you go to medical school for so long and then not use that? But at heart I have remained a doctor, I have only modernized my "form of presentation". Medical cabaret is a "combined preparation". Cabaret is about politics, the comedy faction about sex, men and women and everything below the belt. But all people have a body and a soul, which no one ever talked about, though. With that, I had a topic that appealed to people, they laughed and took something away, they learned something. I call that sustainable comedy: you laugh in the moment and you learn, you have a haha and an aha experience.

Newton vs. Murphy

An example from my stage program "Wunderheiler" (Miracle Healer), in which I try to explain the age-old dispute between "orthodox medicine" and the dazzling world of alternative medicine. A stylistic device for this is personalization, and because I have dabbled a lot in the esoteric field, this was easy for me, because everyone knows a Paul from their own circle of acquaintances.

My friend Paul always orders his parking from the universe, but rides a bike, which is best for him and the universe anyway. We've known each other a long time. We like each other. And we sometimes have very different perspectives on things. The other day we were having breakfast together and his toast fell off the table onto the carpet – butter side down, of course. He immediately exclaimed, "Murphy's Law!" I thought about it for a moment and countered, "Nope Newton!"

There they collided again, our world views. For him, the toast was proof that the stupidest thing possible always happens to him, Murphy's Law. I replied: "Dear Paul, it may well be that you are such a central figure in the universe that there are dark forces that are preoccupied around the clock with how they can make life difficult for you personally. But a basic assumption of scientific thinking is don't assume more things to explain a phenomenon than you need. And for me, all I need in terms of forces is rotation and gravity. The toast has no choice at all, because if it falls from the height of the edge of the table, it can only rotate exactly half a time before it lands on the ground, and therefore it logically lies on the side that was up before."

Paul didn't like being lectured by me and grumbled, "You always with your science." I now really got going, "Paul, that's not my science. It's a process that many people advance in parallel and collaboration, in which you make up theses about the world, test them, and then find the thesis confirmed or reject it."

"Your loss. But what does that have to do with my toast?"

"We can do an experiment to see which one of us is right. You drop the toast, but from twice the height. And if it still lands on the butter side, over and over, then Murphy's Law is true."

Reluctantly, Paul agreed and dropped the toast. What happened? Thanks to earth's gravity and momentum, the slice made a complete turn and landed on its side without butter.

I was so proud and thought Paul was now completely convinced. Fiddlesticks. Paul yelled at me, "Why don't you admit it, you deliberately spread the butter on the wrong side!"

There is much truth in this story. Every human being constructs his own world view, which sometimes fits better, sometimes worse to reality. The idea that the universe is good or bad to us seems easier to bear than the idea that the universe may not care about us personally. When someone comes along who disagrees with and challenges our cherished beliefs, we quickly feel attacked rather than enriched by a point of view. The idea of distinguishing between a sufficient explanation (rotation, gravity) and speculations beyond that (there are additional forces at work) requires a kind of reflection on one's own thinking that is not given to everyone, that is exhausting and, to my deepest conviction, can only be endured with humor.

Gut vs. Head

No human being thinks in only one way. For the vast majority of us, different systems of thinking and believing exist side by side: the intuitive gut feeling and the cool head system that systematically questions things. A gut feeling we have immediately, and often we are right about it, especially if it is an area in which we have a lot of experience. But when it comes to important and far-reaching consequences, it's worth the effort to leave your thinking to more than just your gut.

Here's a recent example about a much more controversial topic: organ donation. How does this become funny? By going for the hidden errors in thinking and exaggerating them.

> The brain is the one organ I would rather be a donor than a recipient of.

The sentence takes a little while to catch on, but it shows that without a brain you are no longer the same. That's why brain death is a prerequisite for organ removal.

A lot of people are scared and think, "I'm gonna get picked apart." No. Not you. It's what's left of you as matter. No brain, no ego. And no sense in selfishness. We'll all be picked apart sooner or later! By worms or heat. What's the difference if you go into the ground or the crematorium with or without a kidney? It's obvious to everyone that you'll never know the difference. We don't say to the undertaker: "Cremation is ok, but please don't make it so hot, Karl-Heinz never tolerated the heat well!"

Stinginess is cool? No. The common good is cool. There's a moment when I can become generous. After I die! It's the ultimate decluttering! I can't take anything with me. I won't need anything.

When it comes to real estate, you can declare personal use. Organ donation means declare personal use! And if you've spent a lifetime sharing all sorts of things on social media – why not share yourself when it matters? I think that's "social." Anyone can write themselves a prescription, for a bit of immortality and fellow humanity. Dissent is encouraged! I believe in life after death – at least in part.

Effectiveness Studies on Cabaret

There is a lot of scientific interest in humor as a means of communication. Because in a world full of funny YouTube films, people's willingness to click through lots of dry stuff on the net has plummeted. The book market is slumping, the live market is booming. And the biggest advantage of a stage program is that no one can switch away and there is no "second screen" in parallel like tablet or mobile phone. Undivided attention is the highest good of an information society.

That's why I agreed to an experiment that, to my knowledge, had never been done before: a control study in cabaret! Students from the Universities of Erfurt and Bielefeld attended four shows of my stage program "Finally!" In two of the shows I performed a piece about organ donation, in the other two I did not. The audience was surveyed three times: before the show, during the intermission, and six weeks afterwards. Using an anonymous code, the data could be analyzed on-site and in the online post-survey. And lo and behold, humor is effective! Among viewers who had heard the stand-up about organ donation, knowledge increased, fear decreased and the willingness to become a donor increased (Völzke et al. 2017).

In another study, my interactive audience action on the topic of "herd immunity" was tested against a classic official website of the Federal Centre for Health Education. Here, too, it was shown that a humorous and original presentation is much better suited to illustrate a complex context than an elaborate animated graphic. The study results from Erfurt and Bielefeld are now being hotly debated at specialist conferences such as "Forum Wissenschaftskommunikation" and in journals. The initiators, such as Cornelia Betsch, Professor of Health Communication at the University of Erfurt, and Florian Fischer from the University of Bielefeld, are also contributing their findings to the "National Action Plan on Health Literacy" and to the recommendations on the role of the media.

Transcend Borders

The media world, health researchers, science journalists, communication professionals and patient organizations often have little contact and exchange. One side knows what the population knows, or more precisely: where the big gaps in health knowledge are. The second side knows how to make good series, magazines or entertainment shows, and the third side knows what patients are actually looking for and what they depend on to navigate the health jungle. It is important that these different worlds and players come together, get to know each other, learn from each other and develop ideas together. Health information is increasingly disseminated through online mass media. However, users are confronted with a great deal of often contradictory and interest-driven information, the origin and quality of which they often find difficult or impossible to assess. Many people have difficulties not only in assessing but also in finding suitable and comprehensible information in the media. This also applies to health-related and medical apps. So far, there are hardly any possibilities to get an overview of these offers or to assess their quality.

The argument that when humor is involved, people don't know what is "serious" and what is not makes people look dumber than they are. I have read with a lot of attention the prophetic book "We amuse ourselves to death" by Neil Postman (1985), and I am not a big fan of quiz shows that ask completely incoherent useless knowledge. However, in my show "Hirschhausen's Quiz of Man" I use the learned structure of an entertainment show to interest several million viewers in prime time for things they wouldn't have tuned in for at first, but they stay tuned. For example, I explained how any layperson can perform CPR – Cardiopulmonary Resuscitation: press on the chest 100 times per minute while keeping the rhythm of the song "Staying alive" in their head. "Dare to do it instead of standing around stupidly," I told my audience.

To date, I have been contacted by five people who have actually dared and successfully resuscitated a person as a result of my broadcast. I think that's wonderful. If the subject of resuscitation were taught in all schools for just two hours a year, we could save a great many lives in this country. They do it that way in Denmark with measurable success. And if we have public television, then surely it is also for the purpose of not just constantly investigating crime stories, but also to teach something. If I can see someone being bumped off every night, I would like to insist that there is also a place for programs that show the opposite: how to bring someone lying motionless on the ground back to life. And if music, humor and an original way of presenting it help – all the better.

What Actually Is Humor?

The Germans are better than their reputation when it comes to humor. Which is not so hard, given their reputation. But what actually is humor?

Humor is a state of mind of cheerful composure, of heartily standing above things. Cynicism, on the other hand, is being above it all without heart and leaves a bitter aftertaste in the end. Humor is also taken seriously in psychotherapy. Humorous stories, twists, haha-experiences can enlighten us like a flash of inspiration in such a way that we suddenly see something differently and some fantasies dissolve in the same second. Many think that if a problem has existed for a long time, it must also take a long time to solve it. That may be, but it doesn't have to be. Humor liberates through the sudden change of perspective.

That's why it's unintentionally hilarious when we find ourselves getting trapped. Like the drunk fumbling in circles around an advertising pillar, shouting, "Help, I'm walled in!" It's obvious to any outsider that all he would have to do to be free is turn around. Only he holds on to the seemingly endless wall and his "worldview."

Why are there jokes and humor all over the world? One of the most convincing theories for me is that it is our mental antidote when we get stuck in a pattern of thinking. People love simple explanations for the phenomena around them, often succumbing to misconceptions about cause and effect. Obviously, humor provides the opportunity to laugh at and correct one's own false assumptions. This is also why a sure sign of any form of ideology is that it always comes across as completely humorless. Those who believe themselves to be in possession of the only truth cannot stand any other perspective.

> A man has lost his way while hiking. Finally he reaches a river and hopes to eventually find a bridge and civilization again. But there is no path, no bridge, nothing. Then he sees a farmer in the field on the other side of the river, tilling his field. Cheerfully he calls over, "Farmer, how do I get to the other side?" The farmer thinks for a while and calls back, "You're already on the other side!"

In humor, contradictions can remain without needing to be resolved. Our mind wants to sort the world, but it is far too complex to be divided into good/evil, right/left, right/wrong. There are three states of the soul in which contradictions are allowed to exist without having to be resolved: the dream, psychosis and humor. One can go mad at the incomprehensibility of life, one can despair of it, or one can laugh at it. Laughter is the healthiest kind and not at all superficial. A great German misunderstanding. With laughter you accept the ambiguity of being, every laugh is a little enlightenment. In all healing cults and religions of the world, humorous stories occur as a vehicle for paradoxes, optimism, understanding and healing. Humor takes the wind out of the sails of fear, even of death.

> A skydiver has jumped out of the plane, pulls the first line – nothing happens. Fortunately, there is still the rescue chute, and so he pulls the second ripcord already slightly panicked. Nothing happens. At full speed he races towards the earth. Suddenly, he can hardly believe his eyes, he sees a man flying towards him from Earth. His salvation? He calls out to him, "Do you repair parachutes?" The other shouts back, "No, just gas lines!"

I told this joke the other day to a man my age who was in palliative care. He had an advanced tumor and was very aware of his situation. He impressed me because, despite his life-shortening illness, he was grateful to be on this ward where he received very loving and competent care. When I told him the joke, he laughed loudly and heartily. For a moment, we were both free, parachute-less in free fall, meeting each other. Those who visit a hospice or palliative care unit for the first time are often surprised to find that there is no sepulchral silence there, but often and gladly laughter or singing, with the energy of "If not now, when?" As George Bernard Shaw said, "Life doesn't stop being funny when we die. No more than it ceases to be serious when we laugh."

US comedian Jerry Seinfeld has given a lot of thought to how his everyday observations have given rise to internationally successful humor. He compares the two poles of a joke to a precipice over which we have to make a mental leap. If the other shore is too close, the attraction of having no ground under our feet for a moment is missing. But if the other cliff is unreachably far, then the excitement is lost and the punchline crashes. In practice, you can tell very well whether you've asked the audience to do too much or too little. A grunt indicates that the punch line was predictable, a laugh that is only sporadic and spread out over time shows that the pennies are dropping one by one and at different rates. A short collective pause and a synchronized burst of laughter are optimal. And that is why good comedy is always a dialogue with the listener, an adaptation to receptivity and speed. Hearing and incorporating the reaction as part of the overall work of art shows the masters of entertainment.

Three Steps to Humor

Humor is like love or football: the people who talk about it the most are not the ones with the most practical experience. There are three things to keep in mind to be guaranteed funny. Unfortunately, none of them are well known.

Economics

Comedy is anti-journalism: the most important thing comes at the end! Comedy has a lot to do with economy. The fewer words used to make the situation clear, the weaker the joke that follows may be. We all resent joke tellers who embellish long and broadly on the back story, only to come up with a tired gag afterward. The same is true of any form of "attention on credit." The longer the build-up and the lead-up to a stunt, the stronger the effect must be, the surprise at the end. Audience and performer have an unwritten agreement: I, the audience, invest time and attention; you, performer, reward me with something that justifies that investment. Thus, with bad jokes, you also feel downright "cheated" out of your precious time. Likewise with a bad movie or crime thriller, if in the end the hours of watching or reading don't pay off.

The Rule of Three

A line is described by two points. And because we think in a straight line, we expect that a third point will also lie on the imaginary line. But the punch line is just not there, but somewhere else entirely. Example.

> A man goes out on the ice early in the morning in the fog to fish. He is about to hack a hole when he hears a deep voice from above: "There are no fish here!" He wonders, thinking he just dreamed it, and keeps hacking. Again the voice comes, "There are no fish here!" This time he is sure it was not his imagination! And very timidly he turns his head toward heaven and asks, "Lord, is it you?" "No," replies the voice, "I am the spokesman for the ice stadium!"

The punch line wouldn't be as funny the first time "There are no fish here" as it was the second time. It wouldn't get any better the sixth time "There are no fish here" but, on the contrary, would stink to high heaven. Hence the rule of three.

Exercise

Now it's your turn. Take your favorite joke and start telling it when the opportunity allows. And opportunities can be created, for example, at the end of a phone call ask if the other person has a minute, you heard a joke the other day, he might like it. Who's going to say no to that? And so, over the course of the day, you have a chance to tell the same joke, first maybe with cues next to the phone, and then later, quite freely, until it becomes part of your own repertoire. And if you have five good jokes in your quiver that you can tell freely, you're already better than most! And so I give you two more jokes from my treasure chest. What you do with them, I leave up to you.

A man enters a train compartment in which two Jews are already sitting. They take turns calling out numbers, whereupon the other laughs uproariously in each case. "14!" "Ow yes, 14, very good! 73!" "73, excellent …" After watching the strange spectacle for a while, the man dares to ask what it was all about. "Quite simple. We both love jokes. But because at some point we'd told each other all the jokes before, we numbered our top 100, and now all we have to do is say the number." The man then calls out, "25!" There is no response from either of them. "Isn't 25 a good joke?" "Yes, but you have to be able to tell it!"

But is there anything left over after the laughter? Is humor, after all, sustainable? Let's try it out with one last joke:

A Dalmatian comes up to the cashier. The cashier asks him, "Do you collect points?"

I love this joke because it's so beautifully pictorial. As always, it assumes something of the listener. First, he has to know that a Dalmatian is a dog with loud dots, and second, that you often hear exactly this stupid question at the checkout. And it demands precision from the narrator: don't try a dachshund when retelling! Only if the dotted dog has been created in the listener's mind's eye in the first sentence can the punchline ignite in the second. The idea of a white, unsullied dog running around collecting points struggles with the absurdity of salespeople constantly and always asking the same question, whether it makes sense or not. And because our minds can't decide between the two interpretations for a "correct" one, there is a logical tension that is discharged in laughter.

If you think humor is unsustainable, I'll bet you today that the next time you're asked at a cash register, "Are you collecting points?" you can't help but think of a Dalmatian and be seized by a strange grin. Only you and the other readers of this book will know the reason. Enjoy it and infect others.

The Author

Dr. Eckart von Hirschhausen (Fig. 10.1) studied medicine and science journalism in Berlin, London and Heidelberg. His specialty: conveying medical content in a humorous way and combining healthy laughter with lasting messages. He has been on the road as a comedian, author and presenter for over 20 years. His current thematic priority is the

Fig. 10.1 Eckart von Hirschhausen – physician and cabaret artist in his doctor's coat. (Photo: Dominik Butzmann)

topic of "climate and health". Most recently, his book "Mensch, Erde! Wir könnten es so schön haben", his search for a better world, was published by dtv.

Reference

Völzke C, von Hirschhausen E, Fischer F (2017) Medizinisches Kabarett als Instrument der Gesundheitskommunikation. Prävention und Gesundheitsförderung. https://doi.org/10.1007/s11553-016-0575-9

11

"Teeth Baring" at acatech

Jaromir Konecny

The German Academy of Science and Engineering (acatech) regularly organizes a "Technik-Derblecken" during Lent. This literally translates into "Teeth Baring", in Bavarian dialect a longstanding tradition for pulling big shot politicans' legs. Following the example of the much-feared ceremony at Munich's Nockherberg "beer temple", a cabaret artist reads the Academy the riot act. In 2018, Jaromir Konecny poked fun at the academic effusions between digitalization and "sector coupling". Here is his "Derblecken" text in English.

I read the Academies' statement on "Social Media and Digital Science Communication": "The Internet and with it the so-called social media have REVOLUTIONIZED

J. Konecny (✉)
Lecturing-Blogging-Performing, Munich, Germany

public, private and political communication as well as science communication!", it says. Just hearing "revolution" gives me panic attacks! ... I grew up in socialist Czechoslovakia!

Of course, I immediately calmed down when I searched the social networks for the German Academy of Science and Engineering: nothing found! Nothing at all! Fortunately, the digital revolution has passed acatech by without causing any damage ...

I found the hashtag #acatech a whole 21 times on Instagram ... Unfortunately, the hashtag didn't lead to the profile of the German Academy of Science and Engineering, but to pictures of Philippine airplanes and lots of beer. But the beer actually had something to do with acatech – it was "Tegernseer Helles" (a popular brew from Lake Tegernsee, Bavaria).

Yes, I know I was reaching too high there. There were only 15 million Germans on Instagram in 2017. I'd better take a look at Facebook, where 30 million Germans are hanging out. Funny enough, I didn't even find an acatech page on Facebook ... Are they building barricades there to ward off the digital revolution?

The academies want to develop a code of conduct for the web and social media. But how can you develop rules for systems that you avoid yourself? That's like me as a Czech trying to teach a German to pronounce words full of umlauts: "Kühlflüssigkeitsüberlaufbehälter", "coolant overflow tank!"

For a few years now, I've had the privilege of moderating the Science Slam for acatech and the Bavarian Academy of Sciences and Humanities. A super show with lots of young people in the audience. I send the bill by e-mail, but that's not enough! You need the original! I still have to sign the invoice and take it to the post office, stand in line there for half an hour and be afraid that there's a mistake in the invoice and I'll have to run to the post office again ... Is that the digital revolution?

You Can't Renew Energy Like Bed Linen

As a German writer, I could only decipher the Academies' statement "Social Media and Digital Science Communication" with the help of the Duden dictionary. For example, the sentence: "The expanded digital communication options take into account increased dialogic and participatory demands", which in German means: "Thanks to Facebook, more and more people can have their say and join in." But, please, tell me! How is science communication supposed to win against "fake news" and pseudoscience when it already conveys even easily communicable messages in such a complicated way that nobody understands them, eh?

In doing so, one could also explain quite complex things to the people in an understandable way. For example, spacetime to a canal digger: "Space-time is when the foreman tells you: 'You have to dig from here to the end of the day!'"

Fortunately, acatech also uses the catchy term "renewable energies" in its writings – which any fool understands, because it's nonsense: renewable energies are fake news that haunts politics – you can't renew energy like bed linen, you can only convert energy.

Technology communication needs a lot of sensitivity! When my mother first visited me in the West in the early 1990s, she pointed to the escalator in the subway station at Marienplatz (Munich Center Square) and said, "You're always raving about how everything works well in the West, but the escalator doesn't work here either." "It works, Mom!", I yelled. "The escalator just stops to save energy, you know? But when someone gets on the platform in front of the stairs, it starts rolling." "Come on!" said my mother, "there's no such thing!" "Yes, there is!" I yelled and jumped on the platform. But the stupid stairs were indeed broken! Since then, my mother only laughed when I tried to explain technology to her.

Even Karl Valentin (legendary Bavarian comedian, 1882–1948) would have enjoyed the acatech paper "Innovation Potentials of Human-Machine Interaction": "Humans and machines are moving closer together" is the first headline. And just below that: "Humans are moving to the center of human-machine interaction!" So man and machine are moving closer together. Like this [*showing a small distance with both hands*]. In the very next sentence, the human moves into the center of it, even though already the human and the machine have moved so close together that the human doesn't even fit in there! The human brain sometimes comes up with things that would leave any machine totally stumped. These are probably also the limits of artificial intelligence. Man wants to play God and make the machine his image.

When I went out for beer with a friend, he said to the ATM, "Please! Can you give me some money?" "The machine doesn't understand you!", I said. "Yes it does!" said my friend. "You're not supposed to piss it off or you won't get anything!" "Bullshit!" said I. "Look!" I put my debit card in the ATM and yelled, "Give me back my money, you bum!" And the machine said, "For security reasons, your card has been confiscated!"

Tons of Tiramisu to Chow Down in Seconds

But the prize for the most comprehensible buzzword should go to the Academies' ESYS working group for "sector coupling". "What do you think of sector coupling?", I asked my ex-wife, and she thought I was making her an immoral offer: sector coupling! Surely the whole academy has been searching for months for a word that would succinctly express how the sectors – "areas" in German – electricity, heat and transport would be more closely linked? Sector coupling!

When it comes to sector coupling, it occurs to me how acatech could couple the acatech sectors "learning machines" and "3D printing": Some AI researchers want to emulate the human brain as in the Hollywood film "Transcendence", i.e. scan the brain precisely and translate these scans and the basic brain functions into a computer program, then recreate the brain on a computer. Acatech could print out the brain reproduced in this way using a 3-D printer – and there we have a 3-D superintelligence made of plastic!

Wouldn't that be emulation of the human brain? I really had to laugh when I read about brain emulation: We've known about the 302 neurons of the *C. elegans* nematode and all 7000 connections between them for about 30 years, and we still haven't been able to recreate this tiny neural system in a computer.

How then are we to emulate the human brain with its 86 billion brain cells and about 100 trillion connections between them? Every single one of those brain cells has at least 1000 to 10,000 connections to other neurons. A transistor on a chip has only a handful of conductor tracks to other transistors. If you want to have your brain fed into a computer to become immortal, you'll have to delay your mortality for a few more centuries until we build such computationally intensive computers – Moore's Law will soon no longer apply, after all. But even if we could build such computers: how do you simulate the quantum leap in the dish (colloquial German for skull) …?

Some techies even want to put implants with data chips in our brains and turn us into super-intelligent cyborgs. But what good is a data implant that slams a few terabytes of information into our brains every second? Our working memory, i.e. our short-term memory, can hold a maximum of eight units of information – eight natural numbers, for example – which are lost again after about 20 seconds. If you had terabytes of information in your brain, it would be as if you had to eat tons of tiramisu in a few seconds. You'd puke after just one kilo!

We also don't need to plug a fiber optic cable into our brain to access the Internet, which some nerds rant about. The human retina can transmit data of almost ten million bits per second and is already optimally connected to the whole brain. What's the big deal?

Some AI experts want to connect two brains with a fiber optic cable and just pump the information across. But they forgot to check with the brain researchers: Even when processing thematically similar information, various parts of our brains fire simultaneously. In order to be exported to another brain, the information or thoughts must first be bundled in a means of communication that works much better between us than the zeros and ones of a fiber optic cable – a development of hundreds of thousands of years: language!

Of course I think acatech's "Learning Systems" platform is great: I'm sure it will find broad support among the people. That's what I told my son. I read to him the acatech statement: "Learning systems will take over the tasks of humans in hostile environments altogether!" "Awesome!" my son said. "School is hostile environments, isn't it? Soon computers will be learning for me, and I won't have to do homework and can just chill! Awesome, that artificial intelligence!"

Threatening Humanity with Cat Pictures

The older ones, on the other hand, are afraid of the AI future: "How are you supposed to emotionally feel in a world of AI machines that you don't understand?" Why not explain to them that they will encounter hundreds of beings every day that you don't understand. Already my ex-wife and I come from different planets – Karin from Venus and I from the Czech Republic.

Why don't science communicators tell people that the new and revolutionary AI cannot become a strong AI, a self-thinking transformer? Many already believe: Soon you'll turn on an AI machine and ask, "Does God exist?" And the AI says, "There he is now!" – Bullshit! Learning systems are just mathematics: a statistical method that can divide data according to established classes and weed out ballast quickly and well – a statistical optimization method that can also be explained by the new mathematical theory of the information bottle neck. No homunculus!

Perhaps the "Learning Systems" project should be given an anti-anxiety section. For many people, it is a horror that someone learns voluntarily. According to the motto: He likes to learn, he must be a complete idiot! That's why the new and revolutionary AI technology *deep learning* neural *networks* is usually translated into German as "tiefe neuronale Netze" (deep neural neworks). The word "learning" is usually left out in German because one does not want to traumatize the Germans with unsympathetic terms like "Lernen".

But what has the new AI learned so far, besides winning at chess and the Japanese game Go? Speech and pattern recognition, but that's wonderful! Crassly good face recognition! Anyone who has used Google Image Search or sorted family photos with Google's Picasa knows what I mean. Dogs and cats can also be recognized super well by AI programs. Because every cat owner posts their cat's picture on the internet, Google's AI is trained with millions of cat pictures. If the Google AI ever wants to threaten humanity, it will flood the entire internet with cat pictures.

The Author

Dr. Jaromir Konecny (Fig. 11.1) is an author, lecturer and blogger on artificial intelligence, poetry slammer (since 1994) and science cabaret artist. He works as a lecturer for Artificial Intelligence at the SRH Fernhochschule and for the SPIEGEL Akademie. His popular science book on Artificial Intelligence "Ist das intelligent oder kann das weg?" (Is that intelligent or is it be discarded?) was published by Verlag Herbig (Langen Müller) in autumn 2020. For the SciLogs of the "Spektrum der Wissenschaft" (Springer Nature) he writes the blog "Gehirn & KI".

Fig. 11.1 Derblecken, "Teeth Baring" 2018 at acatech with Jaromir Konecny. (Photo: acatech)

Since 1992, the chemist with a doctorate has been performing on German-speaking stages, as well as in his old home country of the Czech Republic and elsewhere in Europe. He became a writer and playwright because he likes to make people laugh. His unmistakable trademark: the well-groomed, distinctive Czech accent.

12

Wit and Lightness in Science: The International Perspective

Bruce Lewenstein

A conversation with Bruce Lewenstein, Dean of the Faculty of Science & Technology at the renowned Cornell University, USA. He is an all-rounder and busy world traveller when it comes to science communication. The tenor of the meeting with Wolfgang Chr. Goede in Munich: Science and humor meet shyly.

Science is serious business. Deadly serious for some, says Lewenstein. The search for truth according to the laws of being obviously doesn't tolerate any humor. Not even in the United States, which is supposedly so devoted to entertainment. That's why scientists so often come across to the

B. Lewenstein (✉)
Department of Science & Technology Studies, Cornell University, Ithaca, NY, USA
e-mail: b.lewenstein@cornell.edu

© The Author(s), under exclusive license to Springer-Verlag GmbH, DE, part of Springer Nature 2023
M.-D. Weitze et al. (eds.), *Can Science Be Witty?*,
https://doi.org/10.1007/978-3-662-65753-9_12

world with a stick up their backsides, wooden and nerdy instead of talkative, witty, sparkling. For many, any contact with the public remains awful. And this despite the fact that science communication is slowly opening up to the laws of a democratic society, embracing better comprehensibility, accountability, even good humor … (Peters 2014).

Of Pioneers and Coincidences

Bruce Lewenstein goes to great lengths to set the scene. Two pioneers, gifted performers, broke with deadly seriousness in the USA in the 1950s. The legendary Johnny Carson, who first on radio, then on TV with his "Tonight Show" entertained an audience of many millions for over thirty years, especially with science topics. One of his regular guests was the astrophysicist, exobiologist and writer Carl Sagan.

Among the shining lights of a science that dared to come out of its ivory towers was the ingenious physicist and Nobel Prize winner Richard Feynman, co-developer of the atomic bomb, who made a name for himself in an unconventional way, with wit and fine self-irony. Third in the group tearing down the walls of science was the cartoonist Sidney Harris, who appeared in many specialist journals, even in the intellectual "New Yorker", with whimsical drawings, whose texts then puzzled even the research community. Perhaps it was a good thing that they did not really understand his subtle follies and hidden allusions.

Then the big turnaround. "In the 1970s, a wide gate opened for public science communication, by accident," Lewenstein says. The "New York Times" discovered a new lucrative business model of enriching the paper with a special supplement on weekdays, for example about fashion, and thus acquiring coveted advertisements – only Tuesday was still open, and that's when a newspaper manager fell for science, which had been treated stepmotherly until then.

This move proved to be so profitable that "Science Sections" mushroomed in the US media and soon infected major daily newspapers in Germany with this hype. Another fruit of this marketing strategy was the establishment of special interest magazines on hobbies, sports and science, such as "P.M." in Germany, "Peter Moosleitners interessantes Magazin" (not to be confused with PM, "Popular Mechanics"). Broad, popular science journalism was born, and on both sides of the Atlantic it was in demand for fresh professionals. In Germany, the Robert Bosch Stiftung addressed the shortfall and trained a good hundred science journalists on the fly in the early 1980s.

Accidents and Frankenstein Food

This opening, which science did not want, was compounded by technical accidents that disenchanted research and technology in the eyes of the public. This was long before the Chernobyl nuclear disaster of 1986 and the ensuing chaos of news and information by scientific experts across the board. Lewenstein recalls the British nuclear facility at Sellafield, which had been the subject of criticism since the 1950s with a long chain of mishaps. The grievances and issues, including erroneous forecasts, contaminated milk, destruction of sheep farms, added up to a long catalogue of scientific sins first nailed by British social scientist Brian Wynne (1989).

The hopes of science at the beginning of the last century dissipated in disappointments and fears, which were also increasingly taken up by a critical science journalism. The crash of the space shuttle Challenger in 1986 triggered a future shock, Lewenstein says, comparable to the sinking of the Titanic, and showed how fallible the modern technology gods were.

Partly as a result of this scepticism, the US House of Representatives stopped the construction of the super particle accelerator SCC in 1993, "to which most scientists reacted with sheer incomprehension," Lewenstein recalls. The 12-billion-dollar project showed, says the Cornell scholar, that science has always been, is and will always remain a political issue, everywhere in the world. But most scientists still refuse to see this. The "March for Science" in 2017, in which tens of thousands of people took to the streets for the freedom of science in major German cities and other places around the globe, has done little to change this, despite loud media thunder.

The environmental movement has sown doubts about scientific progress in Europe, more than anywhere else. Genetically modified foods and organisms GMOs were baptized "Frankenstein food" in Great Britain. There, the "Bodmer Report" had already made waves in 1985. The human geneticist Walter Bodmer had advocated a new scientific culture in parliament in order to overcome the deficits of existing, science-centred communication and to build bridges to a new, public understanding of research and social commitment.

Between Governance and Dissemination

This new approach was transferred to other countries and internationalized, especially by the British Council, the cultural institute of the United Kingdom. John Durant of the

London Science Museum and professor for "Public Understanding of Science" (PUS) earned merits in its dissemination. With the turn of the century, the idea of the dialogue model became established (Weitze and Heckl 2016).

This brought the community to a crossroads: Some interpreted the required commitment as a social obligation of science to account for its goals and results, with the greatest possible transparency as well as critical reflection ("governance"). Others understood engagement as taking scientific knowledge to the marketplace, sharing it with citizens, and developing innovative, even emotional forms of science communication for this purpose, participatory and interactive, away from the lectern and towards eye level, and importantly, also with wit and humor ("broad-based dissemination").

Science Festivals and Science Slams

Or can both be combined? Since the dawn of the new century, many new science events have established themselves around these two thrusts. Science festivals and "Science in the City" (large-scale fair-like events) as well as local science slams and FameLabs (in which scientists compete with each other to present their research to the public in the most effective way). Science theatre and bar camps, science clowns and cabaret forms also find their audience, sometimes with thousands of visitors. Formats that promote participation and democracy are booming: from consensus conferences and science debates to citizen inclusion in research projects.

So much for the background, genesis and current state of lightness, wit and humor in science and research, and the debate about them. Are these new forms robust and sustainable? Are they capable of introducing parody and thus self-deprecating, critical reflection into science in the cabaret style, as is standard in political cabaret?

Peer Pressure

All too often humor as such has yet no place among scientists. They tend to feel attacked by humor. Scientists want to stay in their lane, especially since they are under considerable peer pressure. Changes are made only hesitantly, also because science funders and donors prefer a positive image, no leaps in it and certainly no criticism, cautions Lewenstein.

That is one, insistent side of science, on the other, a fresh wind is already blowing. Since the end of the last century, more and more university graduates have settled in journalism, communications and entertainment. However, a cabaret-style parody, such as that attempted by the popular comedian John Oliver on US television's "Last Week Tonight" show, usually features only a pro-science interpretation, reports Lewenstein. But there are many approaches today, all of which help to modernize our scientific culture.

The Author

Bruce V. Lewenstein (Fig. 12.1) is Professor of Science Communication and Dean of the Faculty of Science and Technology Studies at Cornell University. Trained as a historian of science, he also organizes communication training for scientists and is involved in the informal sector of science education as well as citizen science.

Fig. 12.1   Bruce Lewenstein is a lecturer in demand worldwide, here at China's Academy of Sciences. (Photo: Sally Sun)

References

Peters HP (2014) The two cultures: scientists and journalists, not an outdated relationship. Métode Sci Stud J 4:163–169. https://doi.org/10.7203/metode.80.3043

Weitze MD, Heckl WM (2016) Wissenschaftskommunikation – Schlüsselideen, Akteure. Fallbeispiele, Springer, Heidelberg

Wynne B (1989) Sheep farming after Chernobyl: a case study in communicating scientific information. Environ Sci Pol Sustain Dev 31(2):2. https://doi.org/10.1080/0013915 7.1989.9928930

13

"You Don't Understand Science Anyway!"

Constanze Lindner and Wolfgang Chr. Goede

Constanze Lindner is a Munich comedian and performs at the Lach- und Schießgesellschaft (Munich cabaret with national-wide appeal), among other venues. In the interview, she explains why science has been a largely white spot in the cabaret landscape till now. And how it could be filled with color.

Wolfgang Chr. Goede (WCG): When I recently asked a well-known German political cabaret artist after a performance whether he could also imagine science as a topic, he

C. Lindner
Comedy, Cabaret, Wörthsee, Deutschland
e-mail: hello@constanze-lindner.de

W. C. Goede (✉)
Science Facilitation, Munich, Germany

reacted indignantly: "What I propose is completely out of the ordinary." What's the problem?

Constanze Lindner (CL): I think many colleagues have a lot of respect for science.

WCG: Why?

CL: Because it's never really been a cabaret topic before and we don't know much about it. That sets the entry threshold pretty high.

WCG: How are you dealing with that, do you have points of contact and interfaces with science?

CL: The "Big Bang" series on TV with the nerdy Sheldon appeals to me, the wit and irony of how he explains the difference between the moon and an atom, for example. There's also a lot of science in "The Simpsons". In fact, I have something in common with them. I'm Homer's nuclear power plant.

WCG: Hmm, let me guess – maybe because you're always so energetic and radiant on stage?

CL: That's right, I'm a reactor, charged to the tip of every hair with energy that radiates into every niche, lights up every audience member, makes the hall crackle. This image and this state help to overcome many insecurities and self-doubts that gnaw at us stage artists in the difficult art of making people laugh.

WCG: What do you find compelling about science?

CL: There was a teacher once who said, "This table is alive." And told us how it was made of atoms and electrons that were in constant motion, and most of the space in between was just empty space.

WCG: Wouldn't that be a great stage act, to captivate people with a simple table? What it's made of, where the raw material comes from, how different climates shaped the tree, the wood, the piece of furniture. How it is an open book of nature and its changes, including those made by man. The table and its inner life, a product and mirror of our new earth age, the Anthropocene!

CL: Sure, quite feasible. I have created an artificial character, Cordula Brödke, who I sometimes slip into. A young woman with striking braces and very naive. I really have to think about whether I could use that to bring more knowledge to the people. Seriously, this could be a new cabaret challenge for me, with our conversation as an incentive to do so. Overall, though, not easy. Because you have to approach the material without fear, you can't fail with it.

WCG: This caution also because you are a woman?

CL: Evolutionarily, the man is clearly at an advantage, he has more self-confidence. Which, by the way, could be taken apart in a scientific cabaret. Another stumbling block is that science was never made palatable to many of us. Good teachers were the exception. In my childhood and youth, there were such lucky breaks as "Yps" magazine (on basic popular science) or the "Knoff-Hoff-Show" (Know-how) on television. But basically, they said: "You don't understand science anyway!"

WCG: What would be other science topics besides the miracle table?

CL: Profilers who create perpetrator profiles, make DNA traces talk. Absolutely thrilling, with lots of psychology and biochemistry.

WCG: Heaven and universe, man's search for new earths, does that make you tingle?

CL: Not really. In my opinion, man can't really come to terms with infinity and his own tininess in it. But the fact is that our native soil will outlive us. And many people can't come to terms with that either.

WCG: Wouldn't that be a cabaret act, with a hefty dash of sassy satire? How mankind is visibly destroying the foundations of its life. Research, science, technology, the pacesetters of our modern civilization, blowing marching tunes to it while politics scurries impotently about. Even if we get the crises under control, that someday a better adapted creature will reach for the crown of creation and the biped will end like the dino?

CL: Something like that could quickly go wrong. Because it's all too easy for the moral finger to take over. This is something that often bothers me in conventional political cabaret. Despite all the seriousness and drama of the subject matter, the audience must have the freedom to make up its own mind and must never be given a moral cudgel. The balance between seriousness and lightness – no one has mastered it more perfectly than Dieter Hildebrandt in "Scheibenwischer" (German cabaret artist in his TV show).

WCG: That sounds very challenging, how do you achieve this mix on stage?

CL: Two hours of great fun, without judging. And no cuddling. But polarizing is important. "Everybody's darling is everybody's fool!", Franz Josef Strauß knew (former Bavarian Prime Minister). Just bashing, that's not my style.

WCG: Cabaret and science cabaret – how do you see your art in ten years?

CL: We get a lot of new blood from the poetry and slam scene. Despite misanthropes and their claims, German humor has become quite competitive internationally. I see a lot of room for science on the stages. For example, how we can prosper and reach a mature old age thanks to enormous progress in medicine and pharmacy, but also with a view to the controversial ethics involved. I myself am a grateful beneficiary. I suffer from migraines and would not have been able to bear them if the medicines had not helped me. At this point, therefore, hearty thanks: I love science and progress!

The Author

Munich comedian Constanze Lindner (Fig. 13.1) hosts the cult show "Vereinsheim Schwabing" for Bavarian television. She was awarded the Bavarian Cabaret Prize for her stage performance (https://constanze-lindner.de).

Fig. 13.1 Constanze Lindner likes to throw punches. But just hitting and bashing is not her style. (Photo: Martina Bogdahn)

14

Distance, Please!

Hanns J. Neubert

In an interjection, an experienced science journalist urges distance and professionalism. Please don't put scientists on the microphone – just as politicians don't make fun of themselves on the cabaret stage.

It was about time that cabaret took a thorough look at science. Since the times of "Überbrettl" (Berlin), "Elf Scharfrichter" (Munich) and "Jung-Wiener Theater zum lieben Augustin" (Vienna) we know that good political cabaret provides more information and contexts than all news and information broadcasts together (more than theater anyway).

H. J. Neubert (✉)
TELI (German Association of Science Writers),
Hamburg, Germany
e-mail: neubert@sciencecom.eu

© The Author(s), under exclusive license to Springer-Verlag GmbH, DE, part of Springer Nature 2023
M.-D. Weitze et al. (eds.), *Can Science Be Witty?*,
https://doi.org/10.1007/978-3-662-65753-9_14

141

But new paths must be taken. Traditional cabaret, with its left-wing academic cosiness, is a discontinued model. The young generation of cabaret artists, such as Moritz Neumeier or Nico Semsrott, the latter with an audience of millions, "stumble over the political pitfalls of everyday life in which they are personally stuck," as Regina Kusch and Andreas Beckmann put it on National German Radio, Deutschlandfunk. They move away from the big systemic questions and from the idea that, as cabaret artists, they know the score.

Beware of the Exotic Corner

On German television, shows such as "Schlachthof", "Mitternachtsspitzen", "Extra 3", "Die Anstalt" or the "heute-show" have taken up scientific topics again and again. In this respect, I don't really see a gap in the market for extra science cabarets.

By the way, cabaret is more than just laughing about something. It also conveys something other than just humor. And unlike comedy, it also goes much deeper into the substance.

In short, cabaret is satire that loses its edge when it becomes humor or comedy. As Nico Semsrott said, "Joy is just a lack of information."

It would be problematic if scientists themselves took to the stage. After all, political cabaret artists are not politicians. Cabaret artists keep a distance to politics and may/must reveal surprising cross-connections. Science cabaret artists should definitely have this distance – just like science journalists, by the way, who should not be scientists.

A scientific background is of course helpful and enriching. But persons working in science will not be able to be cabaret artists because they are too close to the scientific system. At best, these insiders can be part of a cabaret program, similar to expert contributions in the media that are embedded in an appropriate critical framework.

And let's not forget: Most cabarets have brilliant writers in the background who grind on texts and punchlines and rehearse them with the protagonists.

I actually dread performances in which beginners try out their talent at the open mic. Even cabaret needs to be professionally presented. Let's be honest: The science slams were usually an imposition for all non-slammer colleagues, friends and family members, even if they were well-meaning. It should be similar with cabaret: No one goes to a theater to be told things that don't suit them.

So where are the scientific pitfalls of everyday life and where are people personally stuck in the alternative truths that the scientific experts themselves are constantly broadcasting?

Maybe a little warning in closing:

Science journalists have never really managed to work as normal journalists also in the political, economic, cultural, local and sports fields. That was certainly not entirely their fault but contributed to science journalism often defiantly walling itself off in its exotic corner.

The same could happen to so-called "science cabaret" as a special form of cabaret, especially if one wants to see it as a form of science communication. Then it not only ends up in a cabaret exotic corner, but as outdated, left-wing academic cabaret also loses the charm of real satire.

The Author

Hanns-J. Neubert (Fig. 14.1) is a Hamburg science journalist, former chairman of the German Association of Science Writers TELI e. V., president emeritus EUSJA, European Union of Science Journalists' Associations.

Fig. 14.1 Sailing friend and sailor Neubert: Just like tack and jibe, wit and humor in science have to be learned. (Photo: Ilse Furian)

15

Menu Offering for the Holy Spirit

Martin Puntigam

Martin Puntigam is a member of the Science Busters, Austrian Science Cabaret. He explains that science needs to be packaged in whimsical stories to reach non-specialists. In this set, top science and top humor could become friends.

Why do we have such a hard time to inspire people with science narratives? Quite simply. Because often the narrative is missing. We humans like to hear stories. We've been calibrated to do so for centuries. If you don't offer a story when you tell people about science, you needn't be surprised that your career as a darling of the public stalls. Anyone who only comes up with facts, perhaps even evoking

M. Puntigam (✉)
Science Busters, Vienna, Austria
e-mail: mp@puntigam.at

© The Author(s), under exclusive license to Springer-Verlag GmbH, DE, part of Springer Nature 2023
M.-D. Weitze et al. (eds.), *Can Science Be Witty?*,
https://doi.org/10.1007/978-3-662-65753-9_15

unpleasant memories of unpopular subjects in high school, has already lost.

My memories of physics and chemistry lessons in Austrian secondary school are also unfortunate and pertinent. I remember it well. Physics and chemistry were about as popular in class as a festering coccyx twin: we knew it existed, it must be unpleasant and we certainly didn't want to have anything to do with it!

As boring as a society is in which religions dominate politics and legislation with their fantasies of faith and superstition, one can learn from them that a topic must be packaged in stories.

I understood that early on. As a child, like so many people of my generation, I was a passionate altar boy, so I spent a lot of time with powerful fairy tales, poorly heated houses of worship, and men in drag and women's clothes. It's true that back then, too, the church was all about making contact with aliens who ascended to heaven without a rocket and could turn into bread and wine at will – and back again, even if they were eaten all the time – but back then it had very little to do with natural science. I learned how to combine these stories with science in such a way that you can reach and inspire a large audience from the Science Busters: a science cabaret project that has been trying to prove since 2007 that top science and top humor don't have to be enemies.

Like this: Let's remember Pentecost. A great religious animal story, in which the Holy Spirit plays the main role as a flying edible bird. He comes to earth once a year at Pentecost and with tongues of fire ensures that people suddenly have extremely good foreign language skills. Long before Google was invented.

Questions of Culinary Physics

There is little credible about it from a scientific point of view, except perhaps the tongues of fire, which can give an indication of what is waiting for the Holy Spirit when he re-enters the earth's atmosphere. For if the Holy Spirit comes from heaven to earth as a dove, then at some point he must enter the aerial envelope of our planet. In doing so, he comes into contact with the air molecules, which, as experience has shown, leads to friction. Not spiritually, everything is in the green zone, but from the viewpoint of culinary physics, the following question arises: From what height would he have to start, so that he arrives at the bottom as a well-done pigeon? If heat is already generated during the rendezvous with the air layer, then it should also be used as renewable energy – and with low-temperature cooking, the meat also remains juicier.

Let's say the Holy Ghost jumps from a height of 41 km. That's the current record for human jumps from space. No one has ever tried it from higher up. And now that the marketing value after the virgin leap has been exhausted it's likely to stay that way. And the Holy Ghost, as a feathered sky-diver, has space to himself again. Culinary-wise, however, launching from this height would not be a good idea. A pigeon wouldn't get very warm in such a dive, would be plucked by air friction but would arrive on Earth raw. A delicacy for cats, but not for us humans.

Wishes of the Shooting Stars

However, you should not overdo it with the jump height. With a tenfold increase one comes to clearly higher speeds, but unfortunately nevertheless not to satisfying results.

From an altitude of 400 km a pigeon would reach a speed of about 13,000 km/h before re-entering the atmosphere. That is very fast. There would be no radar penalty for this, because there is no maximum speed for this altitude. But at this speed, a pigeon-shaped object would reach a cooking temperature of about 3600 degrees Celsius in the atmosphere … and would simply burn up.

Not a viable serving suggestion. But if you are lucky you might see the Holy Ghost from Earth as a shooting star. And make a wish. But don't tell anyone what – otherwise the wish has no chance of coming true, as we all know. By the way, scientifically completely unresolved is also the question of whether shooting stars may also wish for something when they see a person.

While a burning pigeon might be too small to make a career as a shooting star, feces make it with ease. However, not those of the dove, but those of people. While the religions common in this country are always hoping for the grace of extraterrestrials, but never get to see one, science and technology have long had stable contact with them. Knowing what they eat and what's left over. "Who are these aliens and why am I just finding out about them now?" you may be asking yourself. But don't worry. You've known about them for a long time. Because the only aliens we humans know about are ourselves. Or rather, those of us who live as astronauts on the International Space Station ISS.

Because transportation to and from space stations is very expensive, as much as possible is recycled. Even feces. Even though there are no immediate neighbors in space, fecal matter is not simply disposed of in the open. Urine is recycled by electrolysis, usually into service water. The resulting feces can be dewatered, i.e. freed from liquid, but not re-used in its entirety. A remainder is pressed and collected in a container. When a supply spaceship flies back to Earth from the ISS, it takes the container with it for a while, ejects it before re-entering the atmosphere, whereupon it burns up. It cannot be ruled out that a couple in love on a clear night silently wishes for eternal love in the face of such a "shooting star".

Back to the Holy Spirit. How do we get him plate-ready? We have to shorten the run-up, like in ski jumping. So that the landing becomes cheerful. From a culinary point of view, the correct altitude is 72 km above the earth's surface. Thereby the pigeon reaches about 2000 km/h on the speedometer. For the cooking process this means: outside 272 degrees Celsius, inside 79 degrees. Perfect! So the Holy Spirit as a pigeon would be crispy on the outside and "well-done" on the inside at his touchdown on earth.

The Author

In the Science Busters team, cabaret artist Martin Puntigam (Fig. 15.1) ensures that the scientists behave reasonably well on stage and do not slip into complete incomprehensibility. He has been awarded twelve prizes for his cabaret programs, books and other projects, including being the first cabaret artist to receive the "Inge Morath Prize for Science Journalism". His engagement with science has finally paid off, as the medical school dropout has been a lecturer at the University of Graz since 2016.

Fig. 15.1 Martin Puntigam in space gear ponders aliens. (Photo: www.pertramer.at)

16

Dictatorship of Stupidity

Jean Pütz

This could also be a scientific carnival speech. Jean Pütz became a prominent face throughout Germany with his "Hobbythek" TV show around technique and his flamboyant moustache. Here, the science journalist humorously sketches a history of mankind that culminates in the abolition of gravity – a post-factual folly, but very serious.

Once upon a time, there was no science and people needed religion to explain everyday life. Whenever an epidemic like the plague or cholera occurred, an evil spirit, like the one Brueghel depicted in his painting, was invented and went about its business from house to house, from window to window. Usually a scapegoat was found for the evil, who was sometimes burned at the stake as a witch. People who

J. Pütz (✉)
Pützmunter Shows, Köln, Germany

© The Author(s), under exclusive license to Springer-Verlag GmbH, DE, part of Springer Nature 2023
M.-D. Weitze et al. (eds.), *Can Science Be Witty?*,
https://doi.org/10.1007/978-3-662-65753-9_16

151

could afford it bought themselves free of their sins by bequeathing money or goods to the church. Sometimes, however, it was the Mother of God who had to serve this purpose as well. Images of her were richly endowed with jewellery or donations, which ultimately also ended up in the church coffers. So far, so bad.

Then came the Enlightenment and philosophers prepared the age of naturalists. At first very slowly, then faster and faster the sciences developed. Technology also benefited from this. It was James Watt to whom we owe the fact that slavery was finally outlawed after a long uprising against it. Posthumously, I award him not only the Nobel Prize for Physics, but also the Nobel Peace Prize. Suddenly slave labor became considerably more expensive than machine labor, and feudal rule chose humanity for economic reasons. Moral philosophers played a minor role in this. Thank technology.

The inventiveness of scientists and technicians seemed unlimited, however with the capacity to destroy everything again: Millions and millions fell victim to colonialism and nationalism, fascism and communism, among others fired up by the new era.

Emotions Remained Stuck in the Stone Age

Personally, I was shaped by the years of National Socialism and its terrible consequences. But afterwards, for 70 years, democracy and peace brought me and my contemporaries an unparalleled streak of happiness. Science was highly respected and also reached the ordinary citizen. Technology made life easier and easier and education ensured that many could benefit and appreciate this.

But early on it became apparent that this explosion of knowledge could no longer reach everyone due to its increasing complexity. In this heyday of knowledge, academics emerged as a kind of parallel society, displaying education as a status symbol. The world became more and more complex and many people, although using the achievements, were left behind. The state tried to counteract this, but it could not eliminate the social differences through education.

But then emerged a phenomenon of social blending. The prosperity of the masses increased more and more and created a new category, which I would like to call "do-gooders", who were guided by their feelings. It was no longer a matter of the scientific principle of cause and effect, but the gut took over the reigns of the brain.

Emotions have such a peculiar meaning. They have not been subject to evolution since the Stone Age. At that time, when Homo sapiens and mankind emerged in Africa, there was no need to make provisions for the future, because in the tropics nature provided everything you needed to live all year round, regardless of the seasons. Life was good, unless the evil neighbour wanted to bash your head in. So there was a sense of belonging to one's clan, but at the same time there was xenophobia. This was essential for survival and has impacted deeply our consciousness.

Because many things remained mystical and could not be explained different religions arose, which built on the primal feelings, in some religions even regarded as proof of God. They and the narrow-minded defensiveness are so deeply entrenched in us that today whole nations can be seduced by autocrats and parties targeting these primitive feelings, as once by the Pied Piper of Hamelin.

This marked the beginning of the age of the post-factual. Here, too, technology, which originally liberated people, played an important role. Never before in human history could the individual obtain so many news. You'd think this would strengthen democracy, but pooh-pooh. There again, a primeval stone-age instinct plays a role, which I will call selective perception. Individuals only pick out the information from the abundance of news and its chaos that happens to fit their world view or stone-age prejudices.

Rational Beings of All Countries, Unite!

They may still contradict people's experience. But the Internet gives them confirmation via the method of Fake News, by now even through algorithmically controlled automated systems. This is the only way to explain that in the country where science was held in the highest esteem a president won the elections who completely negates its findings if they don't suit his political purposes. In order to consolidate his power, he unabashedly spreads fake news and finds unswerving resonance among his group of supporters who are controlled in this way.

But such processes are taking place not only in the USA, but all over the world. Thus, potentates in Venezuela or in Turkey succeeded in making the people believe that they could completely ignore natural and social laws as well as all findings of modern research in order to recruit enough followers to blindly support this dictatorship of stupidity.

In Germany, too, we must be careful that this does not take hold, because both the extreme right and the left are trying to seduce the citizens with promises that can never stand up to reality and science. They act as if they could pull the stars from the sky.

Now imagine that someone from the right or left German political scene comes up with the idea that gravity would be antisocial, because people with higher weight and therefore stronger gravitational force would be disadvantaged. This agitator would then also master the means of mass manipulation, as some politicians may be credited with. To back this up democratically, a referendum would be initiated with the goal: "We abolish gravity!"

Many citizens would be blinded by this and 51% would decide to actually abolish gravity. Then the decision would be handed over to parliament with the mandate to implement it without considering side effects and risks.

You think such a thing is not possible? Then take a look at what happened with Brexit or with Trump and Erdogan or with … Salvation from our post-factual age promises only one remedy: to give reason a chance again – or as I have often called for: Rational beings of all countries, unite! Democratically relying on swarm intelligence is obviously not enough. For this I engage myself with an official page at Facebook and with my daily updated homepage "Science – just arrived!".

It'll save me a psychiatrist in my old age.

The Author

Jean Pütz (Fig. 16.1) is a trained electromechanical engineer who subsequently studied electrical power engineering and communications engineering. He also studied economics and was trained to become a teacher. After a short intermezzo as a senior teacher at the vocational school in Cologne, Pütz became the founder and head of the editorial group "Natural Science and Technology" on WDR television as editor-in-chief, author and presenter. For 30 years he hosted the "Hobbythek" (1974–2004).

Fig. 16.1 Under his magnifying glass, Jean Pütz recognizes developments in society, science and technology that are getting out of hand

Today he is a freelance science journalist, organizes "Pützmunter-Shows", a physical-chemical cabaret, gives lectures at universities, is a multiple award winner for humorous lectures, e.g. at the "Forschologikum Bonn", and "Goldener Narr" (Golden Fool) of the Rhenish Carneval Corporations. For current comments on technology and science, see: www.jean-puetz.net

17

Anecdotes from My Physics Class

Helmut Schleich

The cabaret artist Helmut Scheich tells here with mischievous wit how two cranky physics teachers sharpened his own eye for comic folk theater and parody with their bizarre performances.

To say it first: I graduated from a mathematical-scientific high school in Bavaria. I always had good grades in math, physics and chemistry. My final grade in mathematics was an A!

Why does someone like that become a cabaret artist?

Answer: I was able to follow the science lessons for the most part, but unfortunately it didn't interest me one bit. And yet my mathematics and especially my physics lessons are no minor reason why it drove me to the cabaret.

H. Schleich (✉)
Wortlaut Kulturbüro, Munich, Germany

© The Author(s), under exclusive license to Springer-Verlag GmbH, DE, part of Springer Nature 2023
M.-D. Weitze et al. (eds.), *Can Science Be Witty?*,
https://doi.org/10.1007/978-3-662-65753-9_17

So it wasn't so much the lessons themselves as those who delivered them. Pat and Patachon. With the subtle difference that one (Pat) was not only fat but also tall, the other (Patachon) not only thin but also short.

Pat, in too tight a suit for the Bavarian-baroque corpulence, a gifted pedagogue, however, very old-fashioned with sayings like: "I will teach you in a very brutal way and make you Catholic!" In the lessons, however, he surprised with impressive vividness, so that the number of new members of the Catholic Church was kept within narrow limits.

I was not only amused by the fact that he, with a thinning tonsure, became a red clown in the truest sense of the word when he wanted to pick up a piece of chalk from the floor and his belly got in the way.

Physics Teachers Pat and Patachon

I was not only amused by the way he put his old glasses on his nose more and more diagonally to compensate for the diopters that obviously had like him gotten on in years.

And I was amused not only by the fact that he had apparently already made the entire physics area of the school his biotope to such an extent that he could be found there at (almost) any time of the day or night, as we noticed after the end of a late school party.

It amused me that when he wanted to teach us the direction of the flow of the electric current, he always did so with the extended stinky finger and immediately afterwards threatened the whole class "to teach brutally and to make us Catholic", merely because he was the only one in the classroom who did not know about the double meaning of his pedagogically valuable meant middle finger ("fuck you!"). It was the 1980s. There was still a gap as wide as geological eras between teacher and pupil as far as language and symbolic communication were concerned.

Which gets us to Patachon. He didn't have the pedagogical sovereignty of Pat. Short, but he was wearing a suit that was too big, hectic, always full to the brim with coffee, and correspondingly hectic, chaotic, and blown out of his mind to such an extent that, in the confusion of an unsuccessful experimental set-up next to him, the blackboard with inscrutable scribbling behind him, an exorbitantly high level of caffeine inside him, and resignedly chatting students in front of him, he once ruled over a classmate:

"A Ruah!! (Bavarian for "shut up!"). Otherwise go to the director, he'll give you a blow job." He meant "blow you a march", of course, which in German is to have a serious word with someone. But it was too late for that decisive specification "march". After a moment of incredulous silence in the class, such resounding laughter broke out that poor Patachon seemed as lost as time in a black hole. A physicist no longer understood the world (once again).

He was also the one who once came to sit on the tram on the way to school, directly behind my school mate and me, and quietly cursed to himself. It tore us apart in such a way that a big bang would be a minor detonation in comparison. Today it wouldn't even be noticeable because everyone talks unabashedly to themselves in public anyway. Mostly with barely visible earplugs and talking on the phone. The fact that I still think, when I meet a loudly chattering snotty-nosed brat in Munich-Schwabing, "Wow, a lunatic!" may have something to do with Patachon in his too-big suit and his tram self-talks. Whereas the suits of today's snotty-nosed brats are principally not too big, but too small.

These are just a few anecdotes, but they are to be understood as an appetizer, as an appetizer to a physics class that was held decades ago, that broke the sober boundaries of science at least three times a week and that, in the bizarre performances of its protagonists, gave me a glimpse of what cabaret, when it is good, always is: comic folk theater, parody.

Pat and Patachon were my first subjects of study, my employers in the early stages of school cabaret. And the rising rows of seats in the school's physics hall were possibly already a first glimpse of the theater halls that were to define my life one day.

As far as my physics classes were concerned, I have to be clear: of course science can be funny, especially when it doesn't realize it.

The Author

Helmut Schleich (Fig. 17.1) is one of the most distinctive figures in the German-language cabaret landscape. Television and radio audiences know him primarily as the host of his own political cabaret show "SchleichFernsehen", which runs on Bavarian television and the German public

Fig. 17.1 Helmut Schleich conjures up new ideas for cabaret. (Photo: Martina Bogdhan)

broadcaster ARD, and as a columnist for the satirical weekly review "Angespitzt" (Pointed) on Bavarian radio. Whether on stage, radio or TV – Helmut Schleich takes his audience on adventurous journeys into the depths of the German state of mind and, along the way, shows them the amusing absurdities of everyday life. His solo programs have been awarded the German Cabaret Prize 2013, the Bavarian Cabaret Prize 2015 and the Salzburger Stier 2017, among others.

18

Humor in Knowledge Transfer: Academic Basics and a Workshop Report

Michael Suda

… a plea for a smile finally also in the lecture hall!

Bonn, BMBF (German Federal Ministry for Education and Research). Final event of a research program: 14 PowerPoint presentations in a row. We are the last ones and also we have packed our results into the typical visio- and stereotypes. "There must be a second way around the brain" (Peter Rühmkorf).

Beamer off. Three chairs embody institutions, leaflets and brochures are representative of external communication.

M. Suda (✉)
Lehrstuhl für Wald- und Umweltpolitik, Technische Universität München TUM, Freising, Germany
e-mail: suda@wzw.tum.de

© The Author(s), under exclusive license to Springer-Verlag GmbH, DE, part of Springer Nature 2023
M.-D. Weitze et al. (eds.), *Can Science Be Witty?*,
https://doi.org/10.1007/978-3-662-65753-9_18

So it can be done differently, there is serenity and calmness in the room and the concept of border organizations and border objects has become understandable for everyone. Even years later, the participants will remember it because it was different and because it was "funny". Humor is a key to long-term memory.

They come in big and small, high and low, with and without windows, flat and steep, almost always with uncomfortable seats, sometimes with many of them, but also without students – the lecture halls and seminar rooms at universities. The majority of them are uniformly designed and yet offer great scope for humor to flourish. It is only in recent years that architects have made the discovery that these rooms can also be designed a bit like a stage. These stages serve to impart knowledge.

When is knowledge transfer successful? In the context of a lecture on the awarding of teaching prizes, I recommended to my colleagues and students three elements that can be derived from the conversations about what makes teaching successful:

- It is the tangible enthusiasm of the university lecturer for the material he/she is teaching.
- It is the appreciation of the young people sitting in the audience.
- It is the atmosphere in the lecture hall that is created in the interaction between university lecturers and students.

Humor can certainly contribute to a positive atmosphere, but it can never replace enthusiasm for one's own subject and appreciation of the students. Without this enthusiasm, humor becomes a laughing stock.

Numerous studies on the positive and negative effects of humor in knowledge transfer (a summary can be found in

Wanzer 2002) come to the following conclusions: Humorous teaching personalities are described as more sympathetic, the rigid boundaries between teacher and students are dissolved, which is perceived as positive. On the part of the students, the following positive aspects can be empirically proven:

- increase of awareness
- increase in attendance
- better recall of the subject matter
- better audits
- better evaluation of the teacher
- increase in motivation
- increase creativity
- reduction of anxiety
- stress reduction
- conflict reduction

In addition to these positive aspects, however, the studies also show the limits of humor in teaching situations. Humor can certainly lead to negative effects. This is the case when

- the humor intervention is not related to the topic and is perceived as irrelevant,
- the humor comes at the expense of individual students,
- the humor does not fit the teacher, i.e. it does not seem authentic,
- the teacher makes sarcastic or cynical remarks,
- the teacher acts with mockery or makes sexist remarks and uses stereotypes,
- the students' other sense of humor is ignored or cannot be connected to the students' lifeworld,
- the teacher works with self-mockery and thereby undermines his own authority,
- too much humor puts the seriousness into question.

The studies thus show that there are forms of humor that have an extremely positive effect on the learning atmosphere. However, humor in lecture halls also has its limits. Especially negative forms of humor, which are directed against individuals, are described as distinctly negative.

In principle, humor is a way of creating a different, more relaxed atmosphere or mood in the lecture halls in which the transfer of knowledge is promoted. So it is not about entertaining the students, but rather about getting them interested in the contexts being taught with a few humor interventions. It is also about the self-perception of the teacher, who sometimes puts his own position into perspective and relativizes it, using humor to reduce the distance to the students and acts on an equal footing.

Humor Definitions[1]

Humor is of Latin origin and goes back to the word "umor" which means "moisture" or "fluid". By extension, it refers to bodily fluids, the humors: phlegm (white), blood (red), black and yellow bile. Their respective dominance was considered by the ancient theory of temperaments, which was referred to by the Roman physician Galen in a large-scale work, as the cause of the typological characteristic of the phlegmatic, sanguine, choleric and melancholic.

In theory, the phlegmatic has too much slime in him. Reasonable, governed by high principles, persistent, steadfast and calm, these individuals look at the world. The sanguine has too much blood in him: playful, good-natured, sociable, carefree, hopeful, contented is this type. The choleric has too much yellow bile in him: he is

---

[1] This presentation is based on Titze (n.d.) and the summary of several sources in (2002).

easily excited, self-centered, exhibitionistic, hot-headed, histrionic and active. The melancholic has too much black bile in him and therefore goes through the world anxiously, worriedly, unhappily, suspiciously, seriously, thoughtfully.

"Theory" says that when all four humors are in balance, then a person has a sense of humor. Thus, humor requires a certain balance and serenity.

If we pursue another definition attempt, then we find the following definition in the Duden (Dudenredaktion):

1. ability and willingness to react to certain things cheerfully and calmly,
2. linguistic, artistic or similar expression of an attitude of mind, nature determined by humor,
3. good mood, happy atmosphere.

The first part of the definition describes the ability to react, which, however, requires that things are also perceived accordingly. Humorous people therefore perceive the world in a different way and process this information in a different way. This different state of mind leads to linguistic, mimic or physical actions, at the end of which there is a different mood. If you transfer these aspects to teaching situations, the core is about creating a positive learning atmosphere. Humor is not so much a tool, but rather an expression of the attitude of the university lecturer towards himself and the students.

Humor is divided into two areas of meaning:

1. as the attitude of the university teacher to himself, which affects the image of the self and the world,
2. as a form of communication and interaction with students.

The most difficult gymnastic exercise is still to pull your own leg. (Werner Finck)

All humor begins with no longer taking one's own person seriously. (Hermann Hesse)

Humor as an inner attitude leads to a change in self-perception. This change is accompanied by more optimism and composure. The inner mood also changes the perception of the environment. Fixed points of view are put into perspective by humor. People with a sense of humor are more tolerant and sympathetic. Humor as a social phenomenon changes the mood between people.

Humor is the union of wit and love. (William Makepeace Thackeray)

Humor is – like love – a quality of the heart. (Rudolf Georg Binding)

Reason and genius call forth respect and esteem; wit and humor inspire love and affection. (David Hume)

The quotes link humor with affection, limiting humor to its positive qualities. These definitions predominantly include so-called positive humor and ignore the negative effects of self-deprecating or aggressive humor.

What becomes clear in these definitions is the lack of a clear definition. The term humor is used in many different ways, sometimes as a collective term for everything that is perceived as funny, sometimes differentiated from other forms such as cynicism or sarcasm. For the purposes of this paper, humor will be understood as self-enhancing and socially enhancing humor.

Humor Types

Everyone has a different sense of humor, and what exhilarates one person, another finds embarrassing, silly or ridiculous. In the literature, three types are predominantly described.

The Gelotophobes

On the first type of humor: A broad study (Ruch 2010) comes to the conclusion that in European countries a fluctuating percentage of the population suffers from so-called gelotophobia. This is the fear of being laughed at. While in Denmark it is only 2% of the population, in Great Britain it is 15%. In the Federal Republic of Germany it is 7%. Although there is more to humor than jokes, the reaction to jokes is the most common indicator of humor. In addition to the gelotophobes, two other types are described in the literature (Ruch 2010), to which I would like to add another to complete the picture.

The Flat Humor

This second form of humor usually works with the humiliation of other people or social groups. One makes fun at the expense of others or enjoys their misfortunes. Comedy often uses this form. Whole stadiums are sold out when the protagonists of this form of humor enter the round. People who find this form of "schadenfreude" (malicious joy) funny are certainly common among the audience in lecture halls. Others will reject this form and are more likely to be embarrassed, at best they will protest. This humor likes to work with stereotypes (blondes, civil servants, red necks, police).

> Three blondes were walking. The first said, "My boyfriend gave me a pen, even though I can't write yet." – The second said, "My boyfriend gave me a book even though I can't read yet." – The third said, "My boyfriend gave me a roll-on deodorant even though I don't even have a driver's license yet!"

The Cabaret Humor

This third form of humor focuses on ambiguity or incongruity (Ruch 2010). However, the protagonists of this group always search for the deeper meaning and strive for a resolution. With the resolution, exhilaration then sets in.

> Two cannibals eat a clown.
>   One of them says, "It tastes funny."

The Chaos Humor or Nonsense Humor

In this fourth type of humor, the scenes are chaotic and a clear solution is not obvious (Ruch 2010). The inconsistencies are what make absurdist humor so appealing.

> A man is sitting in a café and observes how the man at the next table, who has just been served a cup of coffee, drinks it up and then eats the cup. He leaves only the handle. Afterwards, the guest pays and leaves the café.

The astonished observer calls the waiter and tells what he has just observed: "The man at the next table drank his coffee and ate the cup. You see, only the handle is left. Isn't that strange?" "That's really strange," the waiter replies, "because the handle is just the best part."

When a group of four people sits at a table, they are unlikely to laugh at the same thing. One shakes his head, the other slaps his thigh, the third leaves the room in horror, the fourth laughs at the absurdity of the situation.

Looking at these types, we can assume that they are also found in different distributions in the lecture halls. What one person might find funny or exhilarating, the others find embarrassing or even an attack on their own person. So we should assume that a humorous lecture – the same is true for forms of interaction or activating teaching – will not appeal equally to all participants. However, my experience shows that a relaxed, humorous atmosphere of exhilaration appeals to the vast majority of the audience.

So everyone has a different sense of humor, and what makes one person laugh, the other perceives as an insult. Humor is thus a very individual matter.

Laughter Is Healthy

"The 'best medicine' is ... nowhere near thoroughly studied. ... children laugh 400 times a day. Adults only 15 times. In the 1950s, people laughed for a total of 18 minutes a day. Today, six. These are the depressing results of laughter research" (Strassmann 2011, p. 33).

Although laughter is healthy, the use of this medicine is decreasing significantly. With worldwide networking, the transmission of mostly bad news has also increased – there's nothing to laugh about! Or can you remember laughing once (except at slips of the tongue!) at the "Tagesschau" or "Heute" (Germany's two principal news outlets on public TV). For health reasons alone, we should do something about this trend.

In gelotology (laughter research) there are a number of findings on what laughter does physiologically, and these studies confirm that "laughter is healthy", but not necessarily for someone who is laughed at. The pioneer of gelotological research is Prof. William F. Fry with the foundation of the Institute for Humor Research (1964).

When someone laughs, it is usually a typical case of a positive emotional state. If a joke turns sour for someone, then a positive emotional state has changed, and this can affect the whole environment. Laughter is an involuntary physical response that is reflexive and preceded by an emotional process. This is in contrast to artificial laughter, which seems contrived because it is cognitively controlled (by the head). Laughter comes from the gut, but the reaction, following humor theories, is triggered by discrepancies in the brain. When we laugh, we emit sounds, breathing changes and the effects on the muscles (abdominal muscle, face) are clearly visible.

What does laughter now do to us? (Anonymous n.d.)

- Men laugh with at least 280, women even with 500 vibrations per second.
- Breathable air is expelled at approximately 100 km/h.
- 300 different muscles are activated (including 18 muscles in the face).
- Heart rate is increasing.
- Blood pressure is rising.
- The fingertips get wet.
- The leg muscles slacken, sometimes even the bladder.

Now we know why some people pee their pants with laughter. After laughter, the organism quickly calms down, blood pressure drops, tension subsides.

The series of experiments by laughter researchers have been continued since Fry's early investigations, and measurability often determined the progress of knowledge. It was also possible to continue the series of measurements "reliably" over longer periods of time, which led to a description of the long-term effects of laughter.

Thus, the following effects were measured (according to Titze n.d.):

- Release and production of stress hormones (cortisol and adrenaline) is reduced
- Promotion of physical regeneration
- Relaxation of the musculature
- Secretion and release of endorphins
- Lowering blood pressure
- Bronchodilatation
- Promote the excretion of cholesterol
- Activation of healthy defense cells
- Proliferation of immunoglobulins and cytokines
- Activation of self-healing powers

US scientists have measured an interesting effect on blood flow in an unusual experimental set-up. The researchers had 20 healthy volunteers watch a clip from a comedy and a war drama at intervals of at least 48 hours. Both before and after the film, the scientists checked the blood flow in the aorta of the upper arm of each study participant by ultrasound. 19 subjects showed accelerated blood flow after watching the comedy. After the war drama, blood flow worsened in 14 of them. Both effects could still be detected at least 30 to 45 min after the end of the film (Huhndorf 2005).

These results provide an indication of the dose of humor that is appropriate in the context of teaching. Too much humor is perceived as rather negative by the students and the teacher loses competence in the eyes of the students. Thus, few interventions are sufficient to create a positive teaching and learning atmosphere.

So laughter – even in the lecture hall – is healthy. What can trigger laughter? What can we use specifically to evoke this reaction?

Humor Theories

> A theory is a conjecture with higher education. (Jimmy Carter, former US President)

So far, a comprehensive theory of the comic is missing (Schwarz 2008). In the theoretical approaches to humor we encounter the so-called classical theories of humor and more modern approaches. Already in antiquity, philosophers, thinkers and researchers have thought about why something is funny and what causes laughter.

Summaries of the theories presented here can be found in Titze and Eschenröder (1998), Sedilek (2009), and Feig (2002), among others.

Here are the classics:

Aggression or Superiority Theory

This theory is about the social and behavioral foundations of humor. This theory sees superiority as the main component of humor. The origins can be traced back to Plato and Aristotle. Plato argues that the stupidity or vices of rather powerless people are laughed at. The main source of laughter is failure, humiliation or the sufferings of other people. Aristotle agrees with this view and sees the origin of laughter in a certain superiority over a person or a quality considered inferior. This laughter is based on schadenfreude (malicious joy) and is close to sarcasm. The theoretical strand also refers to ethnological research and interprets laughter as the further development of animal threatening gestures (showing teeth) or ritualized biting.

In this interpretation, laughter fulfills the following functions:

- Protection from aggression
- Maintaining recognition
- Threat mitigation

Typical examples of the superiority theory are the blonde jokes.

> Two blondes are walking across a bridge – says one to the other, "I want to walk down in the middle too."
>
> Why are blonde jokes always so short? – So that men can understand them too.

Preferably, such jokes are made at the expense of supposedly minorities (Austrians, Bernese, East Frisians), but can also be used as a political weapon. What these approaches have in common is that one feels superior as a non-member of this group. Obviously, resorting to this theoretical approach is extremely successful, if one thinks of the comedy wave that rolls over German living rooms every night like a tsunami. The level turns out accordingly.

"Wit, after all, is always the degradation of another" (Henri Bergson). If one thinks only of this one strand of theory, this quote is true. However, there are other approaches.

Facilitation or Relaxation Theory

This theory approaches the humor phenomenon on the emotional physiological level. Emphasis is placed on the affective-economic side of humor. Laughter reduces tension and is interpreted as a valve for pent-up and thus excess energy. The model is based on the hydraulic nerve energy theory, which states that nerve energy accumulates and is then dissipated by muscle action. Laughter is thus a valve for excess energy.

Sigmund Freud has been cited as the main exponent of the relaxation theory since his writing "The Joke and its Relation to the Subconscious" (1905 [1958]). He holds that one must expend mental energy to suppress hostility and sexual feelings. Any excess is dissipated through laughter. Humor thus serves as an outlet for hostile and sexual feelings and thoughts.

Another approach takes up the Freudian tripartite division of "id", "ego" and "superego". The "superego" is distracted by a joke, the "id" seizes this opportunity and tickles the "ego", which starts laughing. This approach is sometimes called psychoanalytic theory.

The laughter of relief occurs when one has regained security. (This is best observed at the end of Hollywood films – and the soap operas of the previous evening also end with this collective laughter). In horror films, there is usually an "idiot" who provides the tension relief.

Incongruity Theory

It was shown quite early (and by William Fry) that aggression and superiority are not sufficient as an explanation within a humor theory. There was a hint from Aristotle that it might be due to the false expectations of the message receiver. The practiced reader will spontaneously think of the four sides of a message here and open his four ears. All joking aside, the theory comes in knight's armor and makes a compelling contribution to understanding humor responses.

The thoughts of Schopenhauer (1819 [1912]), 200 years ago (an age without television, the internet, the automobile, and the freezer), provide a whole series of exciting aspects to this theoretical approach. He sees the source of humor in a lack of correspondence between our sense perception of things and our abstract knowledge. Different things are viewed under a common concept and referred to by the same words. However, the "biologically older and therefore clearly more developed" senses allow us to separate observation.

When the cave roommate once said, "That was a swallow," it has a different meaning than when the sportswriter says the same phrase today (swallow is a German term for "diving", which in international soccer language is applied when a player fakes a foul).

So there is a consistent reference frame of our perception, which is not sufficiently mapped by the inconsistent oscillation in our brain. This then leads to the humor reaction. This theory is at least able to explain the functioning of many "senseless" jokes.

"I'd like 200 grams of liverwurst, please, of the coarse, fat kind." – "Sorry, she has vocational school today."

Or:

"Why don't ants go to church?" "Because they're insects!"

The wordplay of the examples can be explained very well by this theory.

Laughter is the result of a cognitive short circuit that discharges affectively. In most jokes, a situation is first described that lies within the recipient's horizon of experience and expectation. The punch line is that something unexpected happens. The joke thus consists in a deception of expectation. The recipient thus experiences a disappointment to which he can react in very different ways.

Alternate Theory

This theory is based on the definition of two possible cognitive states. State 1 is characterized by the properties controlled, serious, without excitement and therefore without humor. State 2, on the other hand, is uncontrolled, playful, excited, with humor. There, in Freudian terminology, the "superego" meets the "id" and the "ego" begins to laugh. In literature, this theoretical approach has found little meaning, or shall I say FREU(N)De (German wordplay combining "FREUD" and "friends"). However, numerous cabaret duos successfully work together according to this principle:

- Laurel and Hardy
- Karl Valentin and Liesl Karlstadt (renowned former Munich cabaret duo)
- Loriot and Evelyn Hamann (another more recent pair of famous cabaret artists)
- Jack Lemmon and Walter Matthau
- Bud Spencer and Terrence Hill
- Asterix and Obelix (French comic heroes in constant battle with the Romans)

The range of this theory is rather small, but it provides an important basis for the transfer of knowledge. If the "teacher" leaves state 1 even for a short time and enters state 2 of the "apprentice", he creates a humorous situation, if we follow this theoretical approach. Empirically we can confirm this theory – or maybe it is just a principle – several times. However, it depends on the change.

These opposites are embodied by the duo white clown and "stupid August", also called red clown, the original figures from the circus.

This model is also used in other contexts. On the one hand, the chaotic and irrational is contrasted with the rational and orderly. Children laugh much more often than adults, and they laugh about things that adults do not find funny at all. They are on the side of the irrational. Education transitions this chaotic state into order. Academia is understandably located on the side of order, and the role expectation for university lecturers is predominantly that of the white clown. "Stupid August," who represents chaos, relativizes this order and permanently questions it.

Vera Birkenbihl (2010) worked very successfully with this model. The modern humor theories of Morreall (1983) and Latta (1998) claim to go beyond the classical approaches to cover all situations in which humor occurs. Let us be surprised.

Morreall's Theory

Morreall assumes that laughter is the result of a psychological change. In terms of the incongruity theory in conjunction with the alternation theory, a change from a serious, focused, controlled state to a relaxed, easy-going unfocused state is discovered to be a contradiction. In the case of aggression theory, this change is emotion-related and is associated with

- a sudden surge of positive feeling,
- a negative feeling that fades into the background,
- and the release of a positive feeling.

This approach – without studying it closely – is very reminiscent of "mixing possible" (original German "mischen" sounds like mission, so this reads like "mission possible"). Take a few ingredients from all previous approaches, mix vigorously and claim that the cocktail is a new invention. Shaken, not stirred, please, a new "modern" theory then emerges.

Latta's Theory

This theory claims to be an all-encompassing theory of humor (we will review this claim and reconcile it with our everyday humor experiences). Latta (1998) focuses on laughter, and variable reduction has never done much harm because empirical testing seems within the realm of possibility. "Not laughing" despite a given situation is also described.

According to Latta, the humor process takes place in three phases:

- Initial phase
- Middle transition phase
- Final phase

So far, this is nothing new, but the course of events and all models in this world run according to this proven scheme. Let's start with the translation of the theoretical approach and construct an example.

M. says he is completely relaxed, and his yoga teacher, as well as we, know that there is something wrong with this statement. M. is unrelaxed, and that is the level we are at almost all the time. We are constantly engaged in taking in our environment, making assumptions (The car will stop at the crosswalk, the dog is harmless ...), checking attitudes (She loves me, she loves me not ...), interpreting things or situations (Why did she/he look at me for so long? He/she left the room because ... The sky is blue today because everyone finished their dinner and cleaned their plates). M. is completing tasks or concentrating on something. M. is not relaxed. Relaxation is a myth because M. is awake. So M. is always in the initial phase.

The transition phase is initiated by an external stimulus. The degree of un-relaxation loses its basis for a short time, which leads to M.'s relaxation. This relaxation forms the breeding ground for laughter and the final stage of the humor process is reached. Laughter relaxes M., and as the popular saying goes, "Laughter is healthy."

The model is quite suitable for a description of humor reactions and the mechanism that many a comedian uses, although he/she is probably not even aware of this theoretical approach. The slowly developed joke increases the un-relaxedness, the resolution increases the intensity of the laughter. We recommend further studies that look at different social milieus based on this theory and explore the question of why bad shows need background laughter from a non-existent audience. We hypothesize that program makers have not yet addressed the humor theories presented.

So What Does All This Mean for Teaching?

After all the theory now the question: What can you work with particularly well in lecturing practice?

Not all humor fits into the teaching situation, even if it may work outside (in the entertainment industry). Humor based on power (superiority theory), for example, is certainly not a stylistic device towards participants or students to create a positive atmosphere and achieve relaxation. Word play and other verbal techniques, on the other hand (incongruity theory), can of course be used to great effect in the speaking profession. Playing with two attitudes, two perspectives or two basic states (alternation theory) works particularly well.

Over the years, it has proven successful in my courses to begin the journey inward with an experiential exercise on two original clown figures (this is how it started for me back then, too). The two well-known figures "white clown" and "stupid August" (also "red clown") become models for the role expectations of the lecturer (controlled, serious, knows everything, can do everything better) and the surprising element of a counter-position: childlike, playful, questioning, August shows solidarity with the perspective of the audience, who are expected to accept all the concentrated knowledge of the lecturer. So if one manages to offer/embody both perspectives/poles in alternation in one person, it becomes a cheerful, lively lesson.

Epilogue: Humor in Teaching – Experiences from 52 (Minus One Vacation) University Semesters Looking at Wooden Benches

"What's a lecture?" – "It starts at 8:15 – and if you look at your watch after three hours, it's 8:45." This is not a joke but, following students' accounts, often a reality (… there are other events, though). PowerPoint battles with overloaded information are still commonplace. In the theater or at a science slam, such formats of "knowledge transfer" would cause the audience to leave the room after a short time. We can already hear the outcry of scientific colleagues who point to their paradigm that knowledge transfer has nothing to do with entertainment, audience, let alone humor. The lecturer can neither question his scientific reputation nor his own person. Exactly at this point a chance is missed to gain and keep the most scarce commodity, "the attention" of the audience. Numerous studies (Wanzer 2002) prove the positive effects that humor and a relaxed atmosphere can have.

In his "Philosophy of Magic" Michel-Andino (1994) provides a few central hints, which we have transferred here linguistically to the situation in lecture halls. The lecture is also an artistic activity that is oriented towards the needs of the audience and the goal of which is to make the students' existence somewhat more bearable through laughter, amazement, and the generation of tension and surprise (p. 21). The serious transmission of knowledge is primarily oriented towards the subject matter, the content and the so-called message. The art of entertainment, on the other hand, is oriented first to the audience and thus makes the recipient the actual actor of the event (p. 17). Whoever wants to put the audience in the center of attention from the outset must also make an effort to meet its wishes and needs. He must have a very special relationship with his audience (p. 17 f.).

The special relationship, in my opinion, is to act on an equal footing. This eye level is influenced by different aspects that can be combined with the elements of humor in a miraculous way. All humor starts with not taking oneself (too) seriously. Every sincere scientist is aware of the ephemeral nature of his findings and also knows that every description of this world is a provisional one that passes through a more or less short half-life of memory. The mere presentation of this fact and the relativization of one's own knowledge reduce the vertical distance of eye level.

Appreciation, perhaps even enthusiasm for students are other keys. Those who hate or ignore their audience will look down on students from a higher vantage point, disregarding their needs, motives and interests. Typical forms of humor then are mockery, ridicule, sarcasm, or cynicism, and these expressions have a negative impact on learning outcomes. On the other hand, those who like students, who see themselves as mediators of knowledge, develop other forms of teaching and will constantly rethink their content, perhaps with a smile.

The Author

Michael Suda (Fig. 18.1) holds the Chair of Forest and Environmental Policy at the Technical University of Munich TUM. In his second life, he gives humor courses for university lecturers and performs cabaret on forestry and environmental policy stages.

Fig. 18.1 Michael Suda in professional outfit. (Photo: private)

References

Anonymus (n.d.) Humor und Lachen. https://www.yumpu.com/de/document/read/51844674/humor-und-lachen/. Accessed: 9 Dez 2020

Birkenbihl V (2010) Humor: An Ihrem Lachen soll man Sie erkennen. MFV, München

Dudenredaktion (n.d.) "Humor". https://www.duden.de/rechtschreibung/Humor_Stimmung_Frohsinn. Accessed: 20 Aug 2019

Feig N (2002) Das Phänomen Humor in Lehr- und Lernsituationen der Erwachsenenbildung. Magisterarbeit Erziehungswissenschaft, Heidelberg

Freud S (1958) Der Witz und seine Beziehung zum Unterbewussten. Fischer, Frankfurt (Erstveröffentlichung 1905)

Huhndorf S (2005) Warum Lachen gesund ist. Bild der Wissenschaft https://www.wissenschaft.de/umwelt-natur/warum-lachen-gesund-ist. Accessed: 6 Jan 2020

Latta RL (1998) The basic humor process. A cognitive-shift theory and the case against incongruity. Mouton Gruyter, Berlin

Michel-Andino A (1994) Philosophie des Zauberns: Ein Essay über das Staunen – mit einigen kleinen Täuschungen zur eigenen Erprobung. Krämer, Stuttgart

Morreall J (1983) Taking laughter seriously. State University of New York Press, Albany

Ruch W (2010) Zwischen Lachen und Ausgelachtwerden. Einblicke in die Psychologie des Humors. Forschung Lehre 1(11):14–16

Schopenhauer A (1912) Die Welt als Wille und Vorstellung. Erster Band (Bibliothek der Philosophen 3). Müller, München (Erstveröffentlichung 1819)

Schwarz G (2008) Führen mit Humor – Ein gruppendynamisches Erfolgskonzept. Springer Gabler, Wiesbaden

Sedilek J (2009) Simultandolmetschen und Humor – ein Experiment im Sprachenpaar Deutsch-Englisch. Universität Wien, Masterarbeit

Strassmann B (2011) Unglaublich komisch, DIE ZEIT, No 17, S. 33 http://www.zeit.de/2011/17/Lachforschung?page=. Accessed: 20 Aug 2019

Titze M (n.d.) Humor und Heiterkeit – die wiederentdeckten Therapeutika http://www.michael-titze.de/content/de/texte_de/text_d_06.html. Access: 17 Jan 2020

Titze M, Eschenröder T (1998) Therapeutischer Humor – Grundlagen und Anwendung. Fischer, Frankfurt a. M

Wanzer M (2002) Use of humor in the classroom: the good, the bad, and the not-so funny things that teachers say and do. In: Chesebro JL, McCroskey JC (Hrsg) Communication for teachers. Allyn & Bacon. S, Boston, S 116–126

19

Georg Christoph Lichtenberg: An Early Pioneer of Witty Science

Jürgen Teichmann

Humor in science is not an invention of our days. Already centuries ago it served science communication. Georg Christoph Lichtenberg was particularly successful in humorous science. Witty analogies helped the Göttingen polymath to come up with new experimental ideas.

"The wit is the finder and the mind the observer," Georg Christoph Lichtenberg once pointed out. He was a universal head in a malformed body: professor of pure and applied mathematics in Göttingen since 1770, astronomer, then professor of physics with a famous college in experimental physics, expert in lightning rod and machine questions,

J. Teichmann (✉)
Deutsches Museum, Munich, Germany
e-mail: j.teichmann@deutsches-museum.de

Fig. 19.1 The perfect lightning rod in a mocking letter comment by Lichtenberg: "The iron could contain all kinds of ornaments, e. g. a Jupiter, whom a professor of physics pisses out the lightning." (Photo: Deutsches Museum)

anglophile intellectual, popular science writer, astute mocker and critic, hidden poet between the Enlightenment, Sturm und Drang and incipient Romantic Idealism. He understood wit much more comprehensively than we do today: as the merging of widely separated areas into an original unity, usually with the help of analogies. This wit he classified, even wittily, as "linear," "superficial," or "solid." He used linear jokes for simple pedagogy and communication, for example in his experimental lectures or in his widely distributed popular Göttingen pocket diary (Fig. 19.1).

Of Mnemonic Devices and "Solid Jokes"

A linear joke was also the mnemonic he built for his students to understand the effect of different magnetic poles or different polarity of electricity: unequal poles would attract each other, equal poles would repel each other – that was like man and woman, before marriage they would attract each other and as soon as they were made equal they would only repel each other.

A joke superficial in its sense was actually already sufficiently profound, for example when Lichtenberg emphasized the democratic function of every lightning rod, because it was so easy and cheap to install and worked independently of the social status of the building in question. Only with titles, as common in society, one could construct a difference: "Royal Court Lightning Rod".

By solid wit he understood philosophical insights in aphoristic linguistic form. An example that later became famous reflected, for example, René Descartes' "cogito ergo sum": "Descartes' I think was wrong: *it* thinks, one should say, just as one says it flashes." Sigmund Freud later quoted this in his discovery of the "it" or formally "id" in the human subconscious.

Lichtenberg: Too Popular or Not Popular Enough?

The Romantic poet Jean Paul judged Lichtenberg's literary importance – without knowing the contents of Lichtenberg's then still unpublished aphorism collections, the so-called "Sudelbücher" (waste books):

> Humorous ... is called the noble Lichtenberg, whose four brilliant paradise rivers of wit, irony, whimsy, and sagacity always carry a heavy register-ship of prosaic cargo, so that his splendid comic powers receive their focus only from science and man, not from the poetic spirit.

Today, influenced by objective natural science and technology, we might rather reverse the verdict: too much poetic and humorous fullness, too many flourishes at the instrument of his wit, but too little systematic production of knowledge. In Lichtenberg's approximately 400 works, less than 10% are originally scientific, including the two papers on his most famous publication, Lichtenberg's Figures of 1778 (Glide discharges on insulator surfaces, made visible by scattered powder). 50% of his work is of a popular scientific nature, published in the "Göttinger Taschenkalender" (Diary), which he edited, and in other magazines. The rest is of a literary and other nature. But surely this judgment of Lichtenberg is also wrong, as are many others who have tried to pin down the native Hessian to certain facets of his universal being. The Göttingen professor and writer stood chronologically between Albrecht von Haller, the great physiologist, founder of the Academy of Sciences in Göttingen, famous poet of the Enlightenment, whom Lichtenberg admired without reservation, and between Johann Wolfgang von Goethe, who as a poet in the "Sturm und Drang" published his all-time bestseller "Werthers Leiden" (The Sorrows of Young Werther), which Lichtenberg rejected because of its excess of emotion. Goethe later developed comprehensive conceptions of nature between art, psychology, physiology, physics, for example in his color theory, which Lichtenberg partly welcomed, but partly rejected more strongly – especially with regard to its polemic against the then unassailable Isaac Newton.

Between Experimental Physics and Polemics

An essential leitmotif of Lichtenberg's research as an experimental physicist, but also of his search for knowledge in the rest of the world, was the microcosm-macrocosm analogy, the search for and comparison of the smallest and largest dimensions. The microscope and the telescope served as symbolic instruments for this purpose, also if he himself advanced into the smallest dimensions, especially in the newly emerging theory of electricity. However, aesthetics and pleasure played an essential role in all his endeavours. For example, he described the experiments in which he burned iron springs in oxygen as "the most beautiful I have seen in my life", also as "the most marvellous spectacles", similarly the experiments in which he anticipated modern electric welding. Always there should alternate "a layer utile" and "a layer dulce." Here he was a particularly witty representative of the Baroque age, which sought to bring together aesthetic pleasure and useful enlightenment about the world. "Dulce" (dulcis in latin = "sweet") in Lichtenberg's communication of action and knowledge meant precisely also linguistically and philosophically polished wit. All his aphorisms (he called them "throwaway remarks"), his letters, of course the popular scientific essays, but also his polemical writings are interspersed with it. Wit often became a research tool: witty analogies led to short questions, to conjectures, to new experimental ideas.

But he also often defended traditional knowledge brilliantly: In his defense of the corpuscular theory of light against the wave theory (which did not prevail against the former until after his death in 1799, before undergoing a renewed retrenchment with quantum theory in the early twentieth century), he used striking imagery against the introduction of the ether as the carrier of waves with the statement: It also had been given up to explain "rumbling, crashing, creeping, shining" by "etheric beings", i.e. "ghosts". It is true that Lichtenberg was right in claiming that the wave theory could not yet explain anything experimentally, but his belief in having "the sensuous appearance for us" stood on the same weak feet as the belief of the wave advocates. The common reverence for Newton made him here an exceedingly eloquent advocate of the old theory, brilliant especially where he broke out of scientific reasoning altogether and tried to unhinge his opponent's hypothesis by the means of the joker. This was not a presentiment of modern developments (the special theory of relativity). At that time, a step backwards that is apparent to us today, namely the acceptance of the ether, was necessary for a limited advance, the wave theory of light, to gain a foothold.

Many things, including Lichtenberg's polemics against the substance theory of heat ("Caloricum") of that time, show that his research tool of wit was primarily useful in criticizing the existing development of science. His scepticism towards hypotheses and systems thus gained a great deal of persuasive power. Even so, he remained – all in all – an experimental physicist: one should experiment with ideas, that was another guiding principle for him. For example, he combined physical optics and his keen observation of talented speakers in society: "He could split one single thought that everyone thought simple into seven others like the prism splits sunlight, each more beautiful than the other, and then once gather a multitude of others and bring forth sun whites where others saw nothing but colorful confusion."

From Teaching to Broad Knowledge Transfer

As to Lichtenberg's popular science writing, the boundaries between scientific teaching and broad dissemination of knowledge become blurred. At that time, physics was still part of the philosophical faculty and thus part of the preliminary studies for all three existing faculties (medicine, law, theology). It was held in high public esteem during the Enlightenment and, as experimental physics, did not yet require any complex prior knowledge. Experimental physics got by almost without mathematics. Lichtenberg had assembled from his private fortune an experimental equipment that was unique at that time and he knew how to use it effectively. No wonder that the records of listeners to his lectures give as much of an idea of Lichtenberg's sparkling popular science skills as his articles in the magazines already cited directly demonstrate. And no wonder that as many as 100 students sat in on his lecture (that was about 25% of the entire Göttingen student body!). However, they sometimes slept until it crackled and flashed, as letters tell us.

Lichtenberg was probably the best popular scientist of his time. He possessed

- high scientific understanding,
- broad empathy for the different levels of intelligence and the different worlds of thought of his target groups,
- ingenious abilities to see superordinate connections, i.e. also good knowledge of other disciplinary languages and worlds of thought,
- a literary and poetic talent for the most original possible linguistic or other media-appropriate shaping of this empathy and this seeing,
- great skill in the methodological-pedagogical design of this empathy and seeing.

To the last praise a restriction: Lichtenberg lacked the systematic patience and also an external pressure to carry through larger tasks more long-term. Otherwise he would certainly have written the best and wittiest textbook or non-fiction book on physics of his time. He often had something like that in mind. But nothing came of it. But even so his students, among them later famous ones like Carl Friedrich Gauss, Alexander von Humboldt or Friedrich von Hardenberg (the poet Novalis) admired and enjoyed his knowledge and his art of lecturing.

The Author

Jürgen Teichmann (Fig. 19.2) studied physics in Münster and Munich and, after graduating, the history of science, general history and the sociology of knowledge. He received his doctorate in 1972 and began working at the Deutsches Museum in Munich in 1970, soon as head of educational work (retired since 2006). He habilitated in 1986 on a topic in the history of solid state physics. In 1992 he was project manager of the large astronomy exhibition at the Deutsches Museum. Since 1993 he has been an adjunct professor of the history of science at the Ludwig-Maximilians-Universität LMU in Munich. He has held three visiting professorships (in Hamburg, Pavia, Göttingen). In 2004 he received an honorary doctorate from the University of Gothenburg, Sweden. He has published monographs and articles on the history of science and its didactics. His books for young people turned out also very successful.

Fig. 19.2 Jürgen Teichmann greets us from his study. (Photo: private)

Reference

Braunschweigisches Landesmuseum u. a., Braunschweig, S. 192–200. http://www.lichtenberg.uni-goettingen.de. Accessed: 17 Jan 2020

20

Can the Anthropocene Be Witty? A Science Comic

Helmuth Trischler

Telling stories, presenting science in a graphic way – science comics range from funny to witty and can thus illustrate even unwieldy topics. In them, researchers sometimes take on the guise of Superman …

Does a good comic have to be funny? Not necessarily. Only a portion of comics today offer what this genre, born at the beginning of the twentieth century, programmatically proclaims. Comics have long since broken free of the humorous obligation. As a graphic novel, it is a multimedia narrative in which drawing and text merge into a single entity (McCloud 1993; Knigge 2014).

H. Trischler (✉)
Deutsches Museum, Munich, Germany
e-mail: H.Trischler@deutsches-museum.de

201

If not necessarily funny, the comic should be clever in the sense of being smart and imaginative. The latter is particularly true of scientific non-fiction comics, which have the task of making complex scientific topics vivid in the literal sense by linking text and image sequences in terms of space and content. Since the comic has increasingly emancipated itself from the smell of trivial literature, it is, according to our thesis, particularly suitable for entertainingly presenting and commenting on such a "serious matter" as climate change, biodiversity loss or the Anthropocene as a literary-artistic narrative (Femers-Koch 2018).

Scientific non-fiction comics and science exhibitions have one thing in common: they are slow media that, like slow food, are not geared towards quick consumption, but rather stimulate users discursively and dialogically to new ideas and actions and also to view and question their own position from a different perspective (David et al. 2010). They attach importance to high quality standards in form, content and production as well as in source criticism, classification and weighting of information sources. They appeal directly or synaesthetically to a range of senses and enable disparate information to be combined into complex narratives that operate at multiple levels. They are participatory in that they encourage users to engage intensively with the subject matter, and in this way they promote trust and credibility (Leinfelder et al. 2015).

Communication of the Anthropocene

The media relationship between comics and exhibitions as slow media made it obvious to combine both formats when it came to realizing the world's first major exhibition on the Anthropocene at the Deutsches Museum. The project grew out of a cooperation between the Rachel Carson Center for Environment and Society and the Deutsches Museum, and also involved other partnering institutions, including in particular the House of World Cultures in Berlin, which in 2013 and 2014 with its "Anthropocene Project" also devoted itself entirely to the debate about a new epoch of the Earth shaped by *anthropos* (humankind) (Robin et al. 2014).

What Is the Anthropocene Debate About?

When atmospheric chemist Paul Crutzen and limnologist Eugene F. Stoermer first raised the concept of the Anthropocene in a newsletter of the International Geosphere-Biosphere Programme (IGBP) in 2000, they triggered a scientific debate that quickly gained momentum and is now discussed more widely than almost any other topic (Crutzen and Stoermer 2000). Had the two scientists known this (see Trischler 2016 for the following), they would have published their proposal not in an internal newsletter but in a renowned scientific journal. Crutzen made up for this two years later when, in a one-page article on the geology of humankind in the journal Nature, he succinctly and precisely presented his thesis: humans had become a geological factor to such an extent through their interventions in the Earth that it would require the proclamation of a new geological epoch to reflect this development conceptually. This new age of man, the Anthropocene, had begun with the Industrial Revolution in the late 18th century. Humanity will be the dominant factor in the environment for millennia to come (Crutzen 2002). Since 2009, an interdisciplinary working group, the Anthropocene Working Group, has been examining the scientific evidence for the thesis on behalf of the International Commission on Stratigraphy. In the meantime, it has agreed by a large majority to define the Anthropocene as a formal chrono-stratigraphic unit that began around the middle of the 20th century (Waters et al. 2016).

The discussion about the human age has long since gone beyond the framework of the biological and earth sciences and, as some criticize, has become scientific "pop culture"

(Finney and Edwards 2016). Scholars from many disciplines are now engaged in debates about the Anthropocene that are as intense as they are controversial, and not just in the natural sciences. Remarkably, it is the humanities and social sciences that have entered the discussion about an epoch shaped by humanity on a broad front. This can only be surprising at first glance, since the term itself indicates that nothing less than fundamental anthropological questions are being negotiated here. Moreover, the Anthropocene has long since left the realm of science and is also widely discussed in the media and in public.

The show on display at the Deutsches Museum from 2014 to 2016, "Welcome to the Anthropocene. The Earth in Our Hands" (Figs. 20.1, 20.2 and 20.3) has found many imitators worldwide (Möllers et al. 2015). Currently, several dozen exhibitions of different size and conceptual design around the world deal with the idea of the Anthropocene and communicate it publicly. The Anthropocene is becoming a culturally negotiated issue, blurring the boundaries between science and society. Hundreds of artistic contributions to the Anthropocene that have been realized in recent years show how widespread the concept has become. Art and culture have become drivers of the debate about humans as planetary agents (Trischler 2019).

If one takes the Anthropocene seriously, it blows away established boundaries on numerous levels. The Berlin-based historian of science Jürgen Renn and the science journalist Christian Schwägerl – both committed pioneers of the Anthropocene thesis – argue for nothing less than a radical change in science: in order to truly grasp the consequences of the thesis of humans as geobiological actors and to do justice to the enormous challenges of the Anthropocene, a consistent implementation of both interdisciplinarity and transdisciplinarity in science is required (Renn 2015;

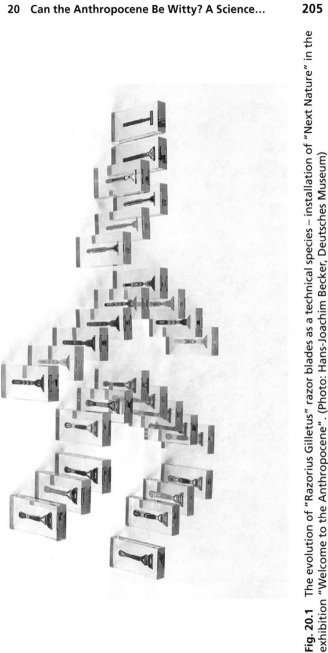

Fig. 20.1 The evolution of "Razorius Gilletus" razor blades as a technical species – installation of "Next Nature" in the exhibition "Welcome to the Anthropocene". (Photo: Hans-Joachim Becker, Deutsches Museum)

Fig. 20.2 A Ward's box terrarium on loan from the Museum für Naturkunde Berlin in the exhibition "Welcome to the Anthropocene". (Photo: Hans-Joachim Becker, Deutsches Museum)

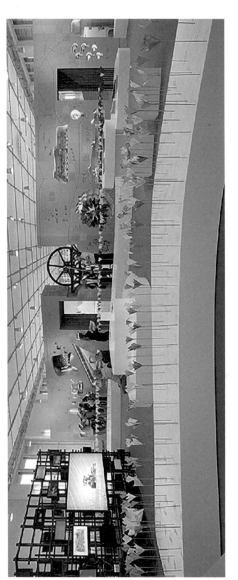

Fig. 20.3 The exhibition "Welcome to the Anthropocene" (Deutsches Museum, 2014) shows the consequences of the rapid development of technology since industrialization. (Photo: Hans-Joachim Becker, Deutsches Museum)

Schwägerl 2013). In other words, anyone who talks about the Anthropocene should not remain silent about the participatory involvement of the public. As the now numerous exhibitions on the Anthropocene have shown, the topic also goes beyond the boundaries of established institutions. To grasp the human being as a planetary actor and to present it in an exhibition in an object-related way requires a multi-perspective approach that links very different collections: With scientific-technical, natural history, ethnological and cultural history collections, only the most important sectors of the still strongly compartmentalized museums' world are mentioned here. This challenge is also evident in the case of the "Welcome to the Anthropocene" exhibition at the Deutsches Museum, which was able to draw on the museum's historically grown collection, but also had to acquire a large number of objects on loan from all over the world (Möllers 2015; Möllers et al. 2015, 2019).

The collections and exhibitions of the Deutsches Museum represent in their entirety the world of science and technology that has grown so extraordinarily rapidly since industrialization and has allowed humanity to become a planetary actor. The geophysicist Peter Haff, who is also a member of the Anthropocene Working Group, assumes that a technosphere now exists as an autonomous, dynamic and global system that can be placed on the same level as the lithosphere, the atmosphere, the hydrosphere and the biosphere. This newly created Earth system, he argues, consists of "the world's large-scale energy and resource extraction systems, power generation and transmission systems, communication, transportation, financial and other networks, governments and bureaucracies, cities, factories, farms and myriad other 'built' systems … all the parts of these systems, including computers, windows, tractors, office memos and humans" (Haff 2014, p. 127). The technosphere, he argues, is not primarily a system created and controlled by humans. Rather, modern humanity is the product of this autonomous system, which operates beyond individual and collective control and imposes its own demands on human behavior.

The provocative thesis of an autonomous technosphere fundamentally contradicts the consensus of the humanities, social sciences and cultural studies that technology is man-made and precisely does not escape human influence. It can – and must – therefore be discussed critically (Trischler and Will 2017, 2019). However, it is helpful in underlining the fundamental importance of technology in and for the Anthropocene. In this perspective, the Deutsches Museum offers a kind of overall view, which confronted the makers of the special exhibition "Welcome to the Anthropocene" with the challenge of linking it to the permanent exhibitions.

Comedy in the Anthropocene

This challenge was attempted to be met with two witty formats. The first format was an "Anthropocene Slam", in which scientists from all over the world were invited not to give a classical lecture, but to create a performance around an object. A selection of these objects was then displayed in a small special exhibition in the special exhibition, which was designed as a multimedia and synaesthetic "Wunderkammer des Anthropozäns" (Cabinet of Curiosities of the Anthropocene) and later also resulted in a brilliant book with the collaboration of the internationally renowned nature photographer Tim Flach (Mitman et al. 2018).

The second format was a factual comic that also revolved around objects. The idea for this came from the geobiologist and member of the Anthropocene Working Group, Reinhold Leinfelder, and the media designer Alexandra Hamann, who had previously worked together to visualize the main report of the German Advisory Council on Global Change (WBGU) "World in Transition – Social Contract for a Great Transformation" from 2011 in a comic (Hamann et al. 2013). Together with the author of this article, they invited Hennig Wagenbreth's illustration class at the Berlin University of the Arts to engage intensively with the pre-selected exhibition objects and translate their reference to the Anthropocene into a visual story. Prior to this, in close exchange with the museum's curators, 30 artefacts from as many different permanent galleries as possible had been identified to mark "milestones on the way to a new Earth epoch", the subtitle of the printed comic anthology (Hamann et al. 2014). In order to achieve a formal recognition value in the exhibition and a uniform graphic design in the accompanying book edition, the comic strips were given a clearly defined format. The drawn stories were to be told in eight panels of equal size each, with only black, white and one special color allowed. This provided a fixed dramaturgical framework, and the uniformity of color enabled recognition of the individual comic strips, which were displayed as banners in the permanent exhibitions in direct spatial connection with the selected original exhibits. The result was an unexpectedly colorful and innovative kaleidoscope of personal narratives that placed science and technology as drivers of the Anthropocene in a very personal, sometimes socially critical context. In keeping with the spirit of Slow Media, the picture stories encouraged museum visitors and comic readers to engage in personal reflection, or as Klaus Töpfer, former Federal Minister for the Environment, Executive Director of the UN Environment Programme and Founding Director of the

Institute for Advanced Sustainability Studies (IASS) in Potsdam, commented, "to think about what consequences our decisions for certain technologies today will have for the development of society and the environment tomorrow" (Hamann et al. 2014, p. 16).

The Dystopian World of the Anthroprocene

In science and technology museums in particular, exhibits are still highly charged with the auratic quality of the original. They embody the supposed objectivity of authentic factual evidence. In contrast, hand-drawn comics make no secret of their subjective interpretation: the strips of the Anthropocene comic "stretched and compressed time periods, simplified scientific findings and historical events, invented protagonists of entire population groups, trivialized pathos, and moved marginal phenomena and coincidence to the center," according to Henning Wagenbreth. For him, who as a graphic mentor granted his students a high degree of freedom, they designed "a collection of pointed narratives for our collective historical consciousness" that make our world more comprehensible "than many a complex scientific work" (Wagenbreth 2014, p. 60).

Quite a few of the strips presented a dystopian world of the Anthropocene, in which a multitude of man-made environmental problems prevail. In her strip on radioactive waste, for example, Nika Korniyenko directly took up the Anthropocene Working Group's debate on radionucleides as primary markers for the new geological epoch (Fig. 20.4) and stated in the accompanying text to her picture story: " Given that the large-scale testing of the atomic bomb in 1950 has left radioactive elements that could send strong, traceable chemical signals into our atmosphere for millennia, the start of the Anthropocene could be set to coincide

NUCLEAR WASTE REPOSITORY --------------------------------

Nuclear power appears to many to be the perfect solution for a CO_2-free and cost-effective energy supply. However, it is only cost-effective for existing power plants that have already been written off, which then entail an increased safety risk - and only if the costs for the disposal of radioactive waste are not taken into account, which to date have been borne exclusively by taxpayers' money. This nuclear waste emits life-threatening radiation over millions of years, with which we burden future generations. Currently it is being considered whether the beginning of the Anthropocene should be combined with the beginning of the nuclear age. Because the from 1960 large-scale nuclear bomb tests behind left such a strong chemical signal of radioactive elements in the atmosphere that they remain detectable worldwide for tens of thousands of years. Much of what we are doing now has effects in geological time - spaces.

N.K. *The impact of human activity on the environment and on our planet cannot be ignored for a long time. My greatest concern is nuclear waste and its incredible and unimaginable consequences, yet greed has fueled doubt and questioned scientific research. I tell my story through the eyes of our innocent fellow inhabitants.*

Fig. 20.4 Permanent Disposal of Nuclear Waste, comic strip by Nika Korniyenko 2014. (Photo: Deutsches Museum) For some people, nuclear energy seems to be the perfect solution for an inexpensive, carbon dioxide-free source of energy. However, it would only be inexpensive if worn down nuclear power reactors are used, which comes with great security risks, and when the costs of disposing nuclear waste, currently at the expense of the tax payers, are not taken into account. This radioactive waste will continue to give off dangerous radiation for millions of years, affecting future generations. Given that the large-scale testing of the atomic bomb in the 1950s has left radioactive elements that could send strong, traceable chemical signals into our atmosphere for millennia, the start of the Anthropocene could be set to coincide with the start of the nuclear age.

Artist's comment: It is no longer possible to ignore the environmental impact of human activity on our planet. I am especially concerned about radioactive waste since despite its surreal and unimaginable consequences, greed has encouraged skepticism and called scientific research into doubt. My story is shown through the eyes of mankind's innocent co-inhabitants on Earth. (https://www.environmentandsociety.org/mml/permanent-disposal-nuclear-waste)

Fig. 20.4 (continued)

with the start of the nuclear age." (Korniyenko 2014, p. 24) It can hardly come as a surprise that a history of nuclear waste is illustrated as a narrative that is funny in the sense of being witty, but saddens readers and viewers rather than making them grin or even smile.

The story of "Super Paul", the discoverer of the ozone hole, on the other hand, is not only shrewd but also funny. The graphic narrative by Martyna Zalalyte (Fig. 20.5) uses the stylistic device of humorous exaggeration. She presents Paul Crutzen as a modern hero who, as Superman, gets to grips with the "evil" chlorofluorocarbons and saves humanity from extinction (Zalalyte 2014).

———————————▶

Fig. 20.5 The Michelson Interferometer for Passive Atmospheric Sounding (MIPAS), comic strip by Martyna Zalalyte 2014. (Photo: Deutsches Museum) "We no longer live in the Holocene, we live in the Anthropocene—in an era shaped by the actions of humankind," says Paul Crutzen, atmospheric chemist and Nobel Prize winner. Evidence of this new geological era of humans can be found in the atmosphere. For instance, the Michelson Interferometer for Passive Atmospheric Sounding, a Fourier transform spectrometer onboard Europe's environmental research satellite ENVISAT, can detect 30 different trace gases in the atmosphere, which could provide information on global warming and the depletion of the ozone layer. In the 1980s, the findings of Crutzen and his team were used as the basis for the Montreal Protocol's ban on the use of chlorofluorocarbons (CFCs), identified as the primary cause for the hole in the ozone layer. Without this ban, the ozone layer, which absorbs most of the sun's UV radiation, would likely be completely depleted within the next 40 years.
Artist's comment: I think my story is gratifying: a scientist discovers a hole in the ozone layer caused by CFCs. Although dire consequences lay in the distant future, he managed to initiate the right steps and prevented the problem from worsening. My story of a modern superhero plays with a hyperbole, which, in this case, is actually the truth. (https://www.environmentandsociety.org/mml/michelson-interferometer-passive-atmospheric-sounding-mipas)

M.Z. *I find my story enjoyable: A scientist discovers the hole in the ozone layer and recognises its connection with CFCs. Although the effects are still far away, he takes the right steps and prevents the problem from spreading further. My tale of a modern day superhero plays with hyperbole, which in this case is true to reality.*

"We no longer live in the Holocene, but in the Anthropocene - a man-made age. " says atmospheric chemist and Nobel laureate Paul Crutzen. As indicators of a new Earth age, Crutzen cites, among other things, the consequences of massive changes to all habitats by humans

since the beginning of industrialization. Further evidence can be found in the atmosphere. With the help of the MIPAS measuring instrument, which is part of the European environmental satellite - Envisat, up to 30 different trace gases can be measured in the atmosphere. They provide information about global warming or the degradation

of ozone. Thanks to Paul Crutzen and his team, chlorofluorocarbons, which they had identified as the main cause of the hole in the ozone layer, were banned at the end of the 1980s. Without this ban, the earth would no longer have a protective ozone layer in about 40 years.

Fig. 20.5 (continued)

Outlook

In his widely read book "Menschenzeit" (Schwägerl 2010), the first popular scientific account of the Anthropocene in the German-speaking world, Christian Schwägerl already pointed out in 2010 that the Anthropocene debate not only links deep geological times, historical times and the present, but at the same time also raises the question of what futures are opened up by a reflected use of science and technology. Anthropocene futures, Schwägerl argues, should draw on a critique of the ways in which technology has been developed and used in recent centuries and decades to point to lines of development that give us the courage to tackle problems that are as complex as they are seemingly intractable, such as climate change, biodiversity loss and global injustice. The future of a "good Anthropocene" imagined by a group of scientists close to the US oil and nuclear energy industries in the form of an "Ecomodern Manifesto", a positive – indeed positivist-technocratic – future that would result solely from an increased use of smart technologies, has rightly been heavily criticized by many critical observers of the Anthropocene debate (Asafu-Adjaye 2015). It has negatively freighted the Anthropocene thesis in the eyes of many. And yet, beyond all critical consideration of the problems caused by humanity as a planetary actor in the Anthropocene, it is important, indeed indispensable, to give new validity to philosopher Ernst Bloch's "principle of hope". Promising stories of how environmental problems have been overcome in the past and present can help us to think creatively and act courageously, ignite hope. Such narratives of "slow hope" are all the more motivating when told with humor and verve (Mauch 2019). The Anthropocene can be communicated in a witty, tongue-in-cheek, light and humorous way. Slow media such as the scientific non-fiction comic are particularly suitable for this.

The Author

Helmuth Trischler (Fig. 20.6) is head of research at the Deutsches Museum, professor of modern and contemporary history and history of technology at LMU Munich, and director of the Rachel Carson Center for Environment

Fig. 20.6 Helmuth Trischler in the "Anthropocene" exhibition of the Deutsches Museum

and Society. In the context of the exhibition "Welcome to the Anthropocene" shown at the Deutsches Museum from 2014 to 2016, he developed a liking for science comics.

References

Asafu-Adjaye J (2015) An Ecomodernist Manifesto. https://the-breakthrough.org/articles/an-ecomodernist-manifesto. Accessed: 19 June 2022

Crutzen PJ (2002) Geology of mankind. Nature 414(51):23

Crutzen PJ, Stoermer EF (2000) The "Anthropocene". Global Change Newslett 41:17–18

David S, Blumtritt J, Köhler B (2010) Das Slow Media Manifest. https://www.slow-media.net/manifest. Accessed: 27 Dec 2019

Femers-Koch S (2018) Comic strips für eine ernste Sache? In: Femers-Koch S, Molthagen-Schnöring S (eds) Textbeispiele in der Wirtschaftskommunikation. Springer, Wiesbaden, pp 147–183

Finney S, Edwards L (2016) The "Anthropocene" epoch: scientific decision or political statement? GSA Today 26(3):4–10. https://doi.org/10.1130/GSATG270A.1

Haff P (2014) Humans and technology in the anthropocene. Six rules. Anthropocene Rev 1(2):126–136

Hamann A, Zea-Schmidt C, Leinfelder R (eds) (2013) Die große Transformation. Klima – Kriegen wir die Kurve? Jacoby & Stuart, Berlin

Hamann A et al (Hrsg) (2014) Anthropozän: 30 Meilensteine auf dem Weg in ein neues Erdzeitalter. Eine Comic-Anthologie. Deutsches Museum, München

Knigge AC (2014) KLONK, BOING, WUSCH!!! Eine Kurzgeschichte des Comics. Aus Politik und Zeitgeschichte 64(33–34):16–22

Korniyenko N (2014) Atommüll-Endlager. In: Hamann A et al (eds) Anthropozän: 30 Meilensteine auf dem Weg in ein neues Erdzeitalter. Eine Comic-Anthologie, Deutsches Museum, München, p 24

Leinfelder R, Hamann A, Kirstein J (2015) Wissenschaftliche Sachcomics. Multimodale Bildsprache, partizipative Wissensgenerierung und raumzeitliche Gestaltungsmöglichkeiten. In: Bredekamp H, Schäffner W (Hrsg) Haare hören – Strukturen wissen – Räume agieren. Berichte aus dem Interdisziplinären Labor "Bild Wissen Gestaltung". transcript, Bielefeld, 45–59

Mauch C (2019) Slow Hope: rethinking ecologies of crisis and fear. RCC Perspect Trans Environ Soc 1:2019. https://doi.org/10.5282/rcc/8556

McCloud S (1993) Understanding comics. The invisible art. William Morrow Paperbacks, Northhampton

Mitman G, Armiero M, Emmett RS (eds) (2018) Future remains: a cabinet of curiosities for the Anthropocene. The University of Chicago Press, Chicago

Möllers N (2015) Zum Konzept der Ausstellung. In: Möllers N, Schwägerl C, Trischler H (Hrsg) Willkommen im Anthropozän. Unsere Verantwortung für die Zukunft der Erde. Deutsches Museum, München, 123–124

Möllers N, Schwägerl C, Trischler H (eds) (2015) Welcome to the Anthropocene. The Earth in Our Hands. Deutsches Museum, München

Möllers N, Keogh L, Trischler H (2019) A new machine in the garden? Staging technospheres in the Anthropocene. In: Rodrigues AD et al (eds) Gardens and human agency in the Anthropocene. Routledge, London, pp 161–179

Renn J (2015) Was wir von Kuschim über die Evolution des Wissens und die Ursprünge des Anthropozäns lernen können. In: Scherer B, Renn J (eds) Das Anthropozän. Zum Stand der Dinge. Matthes & Seitz, Berlin, pp 184–209

Robin L et al (2014) Three galleries of the anthropocene. Anthropocene Rev 1(3):207–224

Schwägerl C (2010) Menschenzeit: Zerstören oder gestalten? Die entscheidende Epoche unseres Planeten. Riemann, München

Schwägerl C (2013) Neurogeology: The Anthropocene's Inspirational Power In: Trischler H (Hrsg) Anthropocene: Exploring the Future of the Age of Humans RCC Perspectives 3. https://doi.org/10.5282/rcc/5603

Trischler H (2016) The Anthropocene – a challenge for the history of science, technology, and the environment. NTM – J Hist Sci Technol Med 24(3):309–335

Trischler H (2019) Kunst als kreativer Treiber der Debatte um das Anthropozän. Kunst und Kirche 1(2019):4–9

Trischler H, Will F (2017) Technosphere, technocene, and the history of technology. J Int Committee Hist Technol 23:1–17

Trischler H, Will F (2019) Die Provokation des Anthropozäns. In: Heßler M, Weber H (Hrsg) Provokationen der Technikgeschichte. Zum Reflexionszwang historischer Forschung. Schöningh, Paderborn, 69–106

Wagenbreth H (2014) Geschichtenzeichner als Geschichtsschreiber. In: Hamann A et al (eds) Anthropozän: 30 Meilensteine auf dem Weg in ein neues Erdzeitalter. Eine Comic-Anthologie, Deutsches Museum, München, p 60

Waters CN et al (2016) The anthropocene is functionally and stratigraphically distinct from the holocene. Science. https://doi.org/10.1126/science.aad2622

Zalalyte M (2014) MIPAS. In: Hamann A et al (Hrsg) Anthropozän: 30 Meilensteine auf dem Weg in ein neues Erdzeitalter. Eine Comic-Anthologie. Deutsches Museum, München, 70–71

21

Science Cabaret: A Script

Marc-Denis Weitze

How does a scientist become a cabaret artist? MDW underwent the experiment on the basis of a burning issue.

Here is his master piece at the Munich Science Slam "10 to the power of 1" for the Tollwood Winter Festival 2019, followed by his reflection on what we may learn from it.

M.-D. Weitze (✉)
TUM School of Social Sciences and Technology,
Technical University of Munich, Munich, Germany
e-mail: weitze@tum.de

Artificial Meat (Lecture Text)

I've been working in new technologies for 30 years. And I've seen a lot: artificial intelligence, artificial photosynthesis ... now we're working on artificial meat. Always something new!

The other day I was invited to the company "No Meat". Nice restaurant they had chosen. And on the table, they placed something: [a picture of a chicken appears on screen] this crispy chicken. Smelled a little funny. At first I didn't know if it was real. Or artificial. So much is possible today. And: This company, "No Meat", works on artificial meat.

They wanted to get their point across as clearly as possible. First we talked about the growing world population – and everyone wants to eat meat, of course.

But why actually artificial meat, went through my mind, it can also be natural. So far it comes from our rural idyll in Lower Saxony: [picture chicken coop] My brother has dozens of these coops, always gets the achievement medals from the Farmers' Association. And masses of EU money anyway. This is pure bioeconomy.

Water consumption went down the last few years by more than half. Nevertheless, everything is clean and tidy. Energy-wise, it's almost self-sustaining. The critters constantly produce heat.

All right, food must be supplied. And medicine. My brother has three veterinarians for his 80,000 chickens, so it must warm everyone's heart at how well the critters are cared for. Now he even wants to hire an animal psychologist. For the creatures' well-being.

It works quite well, the mass production [picture: amount of slaughterings in Germany] Look at this capacity. Every German eats far more than 1000 animals in his life ... the chickens don't even fit the statistics. Here in the hall 350 people, extrapolated ... ah, Tollwood audience, the slide feeds all of us. It's all going through the roof worldwide. Global meat consumption will double by 2050. Indians, Chinese … En masse. Everybody has to drive a car. And everyone wants to eat meat. In India. In China. [Pause].

And half of it is for the bin. You know that yourself, all the leftovers.

[Image Winston Churchill] There's a good quote from Oliver Hardy: "We should escape the absurdity of raising a whole chicken of which we eat only the breast or the wings …" Yeah, sure, there's something to that. And Oliver Hardy, of all people, oops, Winston Churchill has to tell us that. He's absolutely right.

I have talked to my people about this. How can you approach this with the new technologies? We did some research and found "No Meat". The ones with the crispy chicken, they told me before dinner how it can be done: [Graphic: production of in-vitro meat] They go to live animals with a syringe. They take out cells, the animals stay alive for the time being. Then they put the cells into a petri dish. And the cell fibers, cell clusters, grow bigger and bigger in the nutrient solution. The cow on the right is watching. Nice and warm, body temperature, 30 cell divisions take about two months. At some point, the dish will be full. Then you can grow the muscle cells in bioreactors, on scaffolds like this, where the stuff gets into shape. Another three weeks. Then through the meat grinder, minced meat. You can use it directly as a burger [picture burger].

That's what this guy from "No Meat" showed us. The meatball and then the burger, looks delicious! This was all before dinner. I had a big appetite, just wanted to grab the piece, without thinking. And then he quickly pulled it back: "That's an original, costs 300,000 euros," he said, "that's not for eating!" Only five years ago, there was a huge press conference on this new process in London: at least food journalists were allowed to try small shreds: "Not quite as juicy as meat", they wrote – but "basically okay". Food journalist. Another profession.

Kind of a great idea. Just multiply any meat cells, eat them. Chicken and pork. [Pause] Ostrich and kangaroo meat. [Pause] Meatloaf and tofu meat? Why not dogs too – for the Chinese market.

[Image Desktop Myoformer] In Japan they want to offer such a device for the home, a kind of thermomix, in which you put – not the dogs, but only a few cells of the darling, for my sake dog cells, and by the evening the stuff has grown and is ready to fry [pause].

And now you ask: Is anyone in Germany doing this too? We did some more research, and surprise, surprise: the pharmaceutical industry is getting involved. You know, the ones with the pills, Nasivin and so on. They suspect real added value there. The world market will be worth 100 billion US dollars in 2030. They are investing heavily in this, also because of … [image tissue engineering] … yes, with these technologies and the right cells, you can grow all kinds of things. Regenerative medicine. Tissue engineering. Artificial meat. All the same method. Make yourself new nostrils (if you've overdosed on Nasivin). Or new ears … all sorts of body parts ….

I had to tell this my uncle right away. [Pause] He has a sawmill in Lower Franconia [Pause]. He's interested in it. Because he's having his 40th company anniversary soon. He wants to do something nice for the employees. Maybe a surprise dinner? [Pause] We thought about it: How about finger food at the sawmill? Salad with artificial fingertips. It's got to be possible to make a salad out of fake meat. Fingertip salad. And then spice it up with Nasivin squirts!

[Graphic in-vitro meat, see above] … oops, we had that already? Oh no, there was another little detail. They forgot to mention it at "No Meat" and then told us about it during dinner. Then, of all times.

The cow isn't just watching. You need this nutrient solution, the calf serum. It's a little complicated. You got to read it in the fine print. To get this fluid the calf has to be alive, but not yet born. The slaughterhouse receives pregnant cows every now and then. They're sent to the basement, where the butchers go to the unborn calf, suck the blood serum out of its heart until there's none left. Then the serum is in the canister, no longer in the calf [Pause].

"Somehow you can't get it any other way yet," says the boss of "No Meat". I just asked myself, why does he say that over food, over chicken?

[Picture chicken] And then he came out with it [Pause] "Yummy?! Mmmh, our first in-vitro chicken, the prototype! Doesn't taste like veal at all, does it?!" Oh man, then that thing was indeed fake meat – he certainly can't offer the stuff with the calf serum to vegetarians … And if you ask me, the matter at my brother's … [picture chicken coop] … might be nicer after all. And if the Indians and Chinese need something, too, we'll just add a few storeys on top. Enjoy your meal.

What Scientists Learn from Cabaret Artists

When a Scientist Comes to a Cabaret Artist

My first lesson in cabaret. Cabaret artist, musician, artist and coach Ecco Meineke (https://www.ecco-meineke.de/) has agreed to explore with me the question of whether and how my scientific content is suitable to be presented in cabaret. One (preliminary) result is my contribution on artificial meat.

Not a scientific lecture with a few funny pictures and punch lines. But a completely different perspective. In cabaret, we decide, we first define the role of the speaker, establish a context, chat en passant – and finally get to the content.

What role would you like to have? Powerful acting, over-acting, maybe even wearing a bunny costume on stage? Varying language (rhythmic, rap, unfinished sentences, making the audience think along)? Adopting a "deep state," seemingly confused and uncertain, being "creative" with graphics, perhaps playing the myopic expert?

Or rather stay with yourself, the scientist, for example: Intellectually exuberant think-tank employee, runs around with lots of notes, full of file folders. Constantly searching for topics, facts, positions. Thematic overkill, rambling and getting carried away, of course also always up-to-date. "What did I actually want to say …?" – The question is how to navigate through the jungle of material to get to important messages.

The different roles offer room for fantasies and stereotypes: omnipotence, melancholy, excessive demands, cover-ups … Which characters suit me, alias MDW? Loriot? Hanns-Dieter Hüsch? Hagen Rether? Piet Klocke? John Oliver? You can find orientation with these role models or rub up against them.

Artificial Meat Cabaret: The Key Points

The topic "artificial meat" is well suited for cabaret. It concerns people, is topical and has abysses.

We begin the story with a fictional meeting of the protagonist (a think tank employee) with artificial meat factory owners. The status quo of factory farming is presented – not accusingly (which would be so easy), but rather admiringly, until this admiration turns into absurdity (with the animal psychologists).

A corny joke (the confusion of Winston Churchill with Oliver Hardy) is allowed in between.

The sawmill always goes down particularly well. It creates striking images in the minds of the audience. Similarly with the Japanese "Thermomix" vision. A problem of the method (required calf serum) is introduced towards the end, and as a solution (if people don't want to change their eating habits after all) a low-tech solution (construction of several additional levels in the stables so that even more animals can be kept) is finally proposed.

A Guide

As a scientist, what did I take away from the first and subsequent hours of cabaret class? Regardless of the topic and the chosen role primordial is:

- A good start:
  - The first impression is important: Within 10 to 15 seconds you must have connected with your audience, the role must be defined.
  - Once the role is established, it must come to life and remain authentic, even in spontaneous reactions to the audience.
- Rules for the lecture:
  - In slam, as in cabaret (as in real life), the balance between "me" (what message do I want to get across?) and "them" (audience response) has to be balanced. The tension between content and performance style must crackle. The performance must neither be overloaded with content nor ingratiate with cheap jokes.

- All sentences and punchlines must be pre-formulated and in the performance of course seem "spontaneous". Improvisation please only towards the end of a tour, the career – or backstage.
- Punch lines: This is a cabaret, not a lecture hall. People expect punchlines in every third sentence! And please not with a big announcement, but rather "by accident", en passant and as concretely as possible from the story context.
- Pictures: are possible. But these should not be described one-to-one. It is better to create dissonance, the so-called satirical refraction, e.g. to appear surprised oneself and to describe an image ("oops – I've never seen that before …") in an unbiased way and to "read" and interpret it for the audience.
- Speaking of reading: Please, no text on PowerPoint slides, and if so, only single words and numbers. Of course, the audience has to be able to read that … not like in scientific lectures.
- If something goes wrong: Showing weaknesses is okay and makes you likeable.

- Specifics of science and technology topics:

  - In the technology cabaret, we not only pin down debates and controversies, but also develop (new) perspectives from them. As a service to the audience, an opinion's picture is to emerge – even with distorted views.
  - Playing with truths (what is possible?) is more interesting than the truth itself.

- Ending:
    - Showing the absurdity and problematic (again).
- Conclusion: The goal of cabaret must remain in sight:
    - Playing with the feelings and the intellect of the audience.
    - I want people to go home enlightened.

The Author

MDW (Fig. 21.1) is Marc-Denis Weitze (see also editorial team in the chapter To Get Started).

Fig. 21.1 MDW has never had so many listeners at his science talks: at his talk "Artificial Meat" as part of the science slam "10 to the power of 1" at the Tollwood Winter Festival 2019 in Munich, the hall was jammed full – and most of them stayed. (Photo: Magdalena Brunner)

22

Done. Now What?!

Wolfgang Chr. Goede

Enough words have been written, now it's time to practice, experiment, MAKE sense of this. The applause of your audience is worth the sweat. It puts your mind and soul into flights of fancy and joy. All you need is DRIVE!

"Can Science Be Funny?" has become a cookbook full of recipes for dealing with science in many humorous ways. Here, professionals and laypeople, the curious and the experimental folks will find a multitude of valuable tips: Which stylistic devices make scientific presentations creamier, how to make science slams more fiery, which ingredients science cabaret requires, what gives satire its authentic bite.

W. C. Goede (✉)
Science Facilitation, Munich, Germany

© The Author(s), under exclusive license to Springer-Verlag GmbH, DE, part of Springer Nature 2023
M.-D. Weitze et al. (eds.), *Can Science Be Witty?*,
https://doi.org/10.1007/978-3-662-65753-9_22

Humor Recipes

But let's not fool ourselves: Just as we must enjoy cooking in order for recipes to be successful and dishes to be delicious for our guests, humor also requires this enjoyment. This runs like a thread through the chapters collected here. The lightness and playfulness required for this is, however, difficult to convey in a field like science, which in its essence embodies the extreme contrary of wit and humor, namely sobriety and seriousness.

It is precisely this hard-to-crack dialectic that has caused many a person to stumble on their way to a looser presentation of science. We want to encourage them to "Keep at it!" and recommend the "Tegtmeier formula" of our co-author Jürgen Teichmann: that is, to turn a content upside down and provide it with an unexpected negation, as explained using the example of the Magdeburg hemispheres (cf. Chap. 8). This makes the brain trip and us smile in surprise.

This requires no ingenuity, but is pure technique, learnable through repetition and practice, just like writing and arithmetic. A real learning guide with many tasks is "The Comic Toolbox" (Vorhaus 1994). Like many other humor manuals, it comes from the generally funnier Anglo-Saxon culture and would be a low-threshold introduction to the job.

With all our dedication and courage: we can never rule out making fools of ourselves with our humor, just as we sometimes burn ourselves at the stove. Everyone has experienced this themselves when telling a joke that somehow stuck. The line of humor, succeeding or crashing with it is a very fine one, as Peter McGraw and Joel Warner impressively describe after a humor search around the globe in their "The Humor Code" (McGraw and Warner 2014).

The Courage of Self-Mockery

Properly, this book title should have a plural s. Why? According to the authors' fascinating examples, there are dozens, if not thousands, of codes. Regional and cultural differences in the perception of humor complicate it. Berlin laughs at different things than Munich and what the Argentinian smiles at, the Dane finds instructive and stiff. For example, the cartoons of Mafalda, a smart aleck who philosophizes about the badness of the world, with which the most renowned daily newspapers in Latin America adorn themselves.

"And you find that funny?" – With this question Henri Nannen, the legendary founder of the Stern magazine, brought tears to the eyes of Germany's most talented cartoonists at the weekly editorial deadline. With so many imponderables, we normal people prefer to play it safe and resort to the serious boring act. The fear of embarrassing ourselves with humor so that it turns on us as mockery runs deep in all of us, but: self-irony and the willingness not to take ourselves so seriously help us get over stumbling beginnings. All who publicly laugh at themselves and their mistakes immediately find fellow laughers, cut their suffering in half and prove that they have grasped the basic form of humor and are on the right track.

Jesters as an Early Warning System

But whoops, isn't there another stumbling block, a really huge one? Fundamentally, scientific humor also raises a moral question, namely that of our dealings with and respect for authorities, especially when they – like scientists – operate in the higher spheres, so to speak, in search for truth. Are we allowed to make fun of them so easily?

A look at history and the role of jugglery in the absolutist power system of the Middle Ages helps to answer this question. Its rulers were the epitome of truth, but: every prince who thought anything of himself afforded himself a court jester. With jokes, comedy and humor, they provided amusement and entertainment at court and for their subjects, often at the expense of their employer. Court jesters acted as a social outlet, with laughter as ingenious means of taking the edge off oppositional forces, pacifying fermenting popular souls, and moreover they were a rather sophisticated early-warning system.

Above all, the court jester was the only one who was allowed to hold up a mirror to the prince or ruler, because no one in the court would ever have dared to say a word of criticism to the man at the top. In the mirror of jokes and innuendos, the superior could see how he was received, where he had deficits, whether he was already making a pitiful naked idiot like in "The Emperor's New Clothes". The jester was critic and opposition, advisor and reformer, all in one – long before the Enlightenment, the separation of powers, the triumph of the press.

Court jesters and jugglers have been handed down to us in many historical guises and we encounter them in the harlequin as well as in the clown. They live on in the card game as jokers and "wild cards" who have the power to do anything. Figures such as Punch and Judy, buffoons and clowns of all walks and cultures, last but not least the immortal cinema heroes of the silent movie era are related to them. The most famous court jester and at the same time the role model for all his colleagues until today is Till Eulenspiegel.

The writer Daniel Kehlmann has brought him to life in his novel "Tyll" and embedded him in the Thirty Years' War

(Kehlmann 2019). By means of the craziest antics and weirdest acts, with which he skilfully subverted the protocols of the time, Tyll survived the ludicrous slaughter of men. He eased and resolved conflicts, made powerful people smile, gave them a lesson in thinking again and again with fine satirical barbs.

This tried-and-tested, centuries-old cultural tradition is also well suited to our future, which is increasingly driven by science and technology. Many of the young academics who do not want to get into research and the regular job market could find a home here. It could also be a new home for (science) journalists who no longer find a place in the shrinking print sector. Nor do the masses of media students need to be thrown out into unemployment. As the contributions in this book show, the two main tasks of journalism, information and criticism, can also be tackled with humor and wit – possibly even more effectively than with the stylistic means used to date.

Profession with a Future: Science Joker!

In medicine and health, psychosomatics and mental wellness around burn-out, depression and anxiety alone, there would be a multitude of attractive applications (see box). Other areas of research and everyday life with a high demand for more wit and comedy are pedagogy and education. And why do Jane and John Doe understand so little legal jargon and get so easily tangled in the jungle of paragraphs without professional and often costly advice?

Burn-Out, Anxiety and Trauma Cabaret

The laughter of the hospital clowns is supposed to dispel gloom, promote the healing process, blow away fear. In more and more pediatrician's offices, a funny rag doll greets the little patients, to which the doctor likes to refer to relax them before injecting them with a shot: "Lotta already got one today, too." The idea is currently being taken further in a congenial and co-creative way. After all, health has an increasingly important psycho-mental component that needs to be served with wit and humor.

Depression, anxiety, burn-out are considered new widespread diseases in the OECD countries. As an artistic novelty in this risk zone, a burn-out cabaret debuted in Munich's Gasteig in summer 2018. It humorously skewered booming mental illnesses and provided many tips on how working people could resist them. The performance was organized by a medical doctor and founder of a burn-out clinic that incorporates creative cabaret elements into therapy, and this premiere particularly appealed to the healing professions. Hand puppets, familiar from Punch and Judy or the "Muppet Show", are also used. These can be used to play through psychologically stressful scenes, including traumatic situations, in front of small or large audiences, and to garnish them with a great deal of humor, for example by a Courage Punch and a Fear (making) Crocodile: funny and cheerful as in puppet shows, with a great learning factor.

Educators as well as self-help initiatives could benefit from this for their work. In addition, puppet dialogues between teachers and students, professional academic experts and lay experts by experience, decision-makers and critics could enliven every teaching situation, lecture and debate and add a new dimension to the culture of learning, healing and debate (http://netzwerk-gemeinsinn.org/der-angst-kasper).

That's right, without a coach, lawyers themselves can't get out of their unwieldy speech. That's why the first clowns take to the US lecture halls, learned lawyers who use wit and humor to train and educate their colleagues in breaking down complicated matter for their clients. Before we know it, it may even be possible to become a science joker – as newly offered academic courses to supplement scientific-technical subjects.

Not really a joke, just the logical conclusion to this book. Welcome to the brave new world of smiles!

The Author

About the author (Fig. 22.1): see editorial team in Chap. 1.

Fig. 22.1 At the Munich Science Days 2017, Wolfgang Goede played a sex robot and slammed about the creative relationship between artificial intelligence and love. (Image: acatech)

References

Kehlmann D (2019) Tyll. Rowohlt, Reinbek

McGraw P, Warner J (2014) The humor code. Simon & Schuster Paperbacks, New York

Vorhaus J (1994) The comic toolbox. Silman-James Press, West Hollywood

Postscript: Corona Cabaret Critique

The book manuscript was ready in early 2020. Then the pandemic hit. No more fun?

Since then, science has been brilliant, sequencing the viral genome in record time, establishing tests using PCR, launching numerous vaccine developments. It assesses risks in classrooms and supermarkets, churches and football stadiums, opera houses and karaoke bars.

General trust in science and research has risen significantly against the backdrop of the Corona Covid-19 pandemic, notes the "Science Barometer" in April 2020. German science communicators like Christian Drosten and Mai Thi Nguyen-Kim are showered with awards. Politicians are delighted: "It is the hour of the science explainers" (ED Rossmannn, S Kaufmann in WELT online, 02.06.2020).

Are these now stellar moments in science and science communication? Is information critically classified and provided with context? Does science advise policy-makers and the public in a comprehensible, reliable manner and with the necessary independence and distance? Are doubt, scepticism and criticism unrestricted as basic elements of science? Would humor perhaps be appropriate right now?

Or should we take the whole incident as an example of how science takes itself far too seriously, politics desperately seeks orientation, and the media report uncritically and in "informational idleness" (according to media researcher Michael Haller)?

Much was heard from science on the lockdown, but did it really have anything to say? Science still (as of October 2020) provides: no reliable rapid test. No therapy. No vaccination. We have to deal with the virus in the twenty-first century like we dealt with the plague in the Middle Ages: masks, distance, quarantine.

The political decisions on the Corona measures in Germany had to pragmatically follow common sense: A statement by the National Academy of Sciences Leopoldina in April 2020 sought to describe strategies that could contribute to a gradual return to social normality – but was no more than the expression of opinion by a group of scientists. In the public and media, the measures were widely accepted as having no alternative in the first few months, with little criticism. Media researcher Stephan Russ-Mohl noted a paralysis of shock in the early stages of Corona's coverage: "Criticism wasn't much in demand then." (Media Week, 4/23/2020). The public even witnessed politicians and journalists sweepingly denigrating people who demonstrated against current policies. And the moderators of "Die Anstalt" (German satirical TV show, broadcast of 02.06.2020) indulged in a mockery of the "covidiots", not even considering that the improvised Corona measures themselves have long since provided endless cabaret material.

There is another way, and this book pleads for these forms of communication:

Sascha Lobo clarified in October 2020 in his satire overview: "Mask obligation prevails only on even days in streets with more than nine letters weekdays between 5.45 a.m. and 10.08 p.m." (Spiegel "Netzwelt" 14.10.2020).

The cabaret artist Pigor already sang in April "jetzt! jetzt! jetzt!" ("now! now! now!") https://www.pigor.de/songs-a-z/ "We should talk about the success of business models – Which charge their follow-up costs to the general public. And it's legitimate to use Corona as an occasion – Because the next crisis is bound to come. We shouldn't miss the moment now – When politicians let scientists tell them what to do."

So now it's time to take a critical and cabaret approach to the Corona crisis. The pandemic creates some funny situations, just with reference to the masks and their use, sense and nonsense:

- A small virus cripples our world, steeped in science and technology, the pinnacle of civilization. For months, scraps of cloth are our only protection.
- We all have to wear "mouth-nose coverings" on public transport. On the Munich subway, it's been a long time the rule: "Dogs that can endanger passengers must wear a muzzle." Enters a man with an attack dog on the subway. Both look dangerous. An elderly lady just slips her mask under her nose. How do you think the story goes …?
- The months-long debate about the pros and cons of mouth-nose protection and side effects can be traced on quarks.de; a search with several dozen references to scientific sources ends up like "Hornberger Schießen" (German idiom for "no results whatsoever"): Nothing known for sure.

Science should now not only go in search of vaccines and prevention, but also find its wit again.

Autumn 2020
The editors

Printed in the United States
by Baker & Taylor Publisher Services